MATHEMATICS RESEARCH DEVELOPMENTS

A CLOSER LOOK AT THE DIFFUSION EQUATION

MATHEMATICS RESEARCH DEVELOPMENTS

Additional books and e-books in this series can be found on Nova's website under the Series tab.

MATHEMATICS RESEARCH DEVELOPMENTS

A CLOSER LOOK AT THE DIFFUSION EQUATION

JORDAN HRISTOV
EDITOR

Copyright © 2020 by Nova Science Publishers, Inc.

All rights reserved. No part of this book may be reproduced, stored in a retrieval system or transmitted in any form or by any means: electronic, electrostatic, magnetic, tape, mechanical photocopying, recording or otherwise without the written permission of the Publisher.

We have partnered with Copyright Clearance Center to make it easy for you to obtain permissions to reuse content from this publication. Simply navigate to this publication's page on Nova's website and locate the "Get Permission" button below the title description. This button is linked directly to the title's permission page on copyright.com. Alternatively, you can visit copyright.com and search by title, ISBN, or ISSN.

For further questions about using the service on copyright.com, please contact:
Copyright Clearance Center
Phone: +1-(978) 750-8400 Fax: +1-(978) 750-4470 E-mail: info@copyright.com

NOTICE TO THE READER

The Publisher has taken reasonable care in the preparation of this book, but makes no expressed or implied warranty of any kind and assumes no responsibility for any errors or omissions. No liability is assumed for incidental or consequential damages in connection with or arising out of information contained in this book. The Publisher shall not be liable for any special, consequential, or exemplary damages resulting, in whole or in part, from the readers' use of, or reliance upon, this material. Any parts of this book based on government reports are so indicated and copyright is claimed for those parts to the extent applicable to compilations of such works.

Independent verification should be sought for any data, advice or recommendations contained in this book. In addition, no responsibility is assumed by the Publisher for any injury and/or damage to persons or property arising from any methods, products, instructions, ideas or otherwise contained in this publication.

This publication is designed to provide accurate and authoritative information with regard to the subject matter covered herein. It is sold with the clear understanding that the Publisher is not engaged in rendering legal or any other professional services. If legal or any other expert assistance is required, the services of a competent person should be sought. FROM A DECLARATION OF PARTICIPANTS JOINTLY ADOPTED BY A COMMITTEE OF THE AMERICAN BAR ASSOCIATION AND A COMMITTEE OF PUBLISHERS.

Additional color graphics may be available in the e-book version of this book.

Library of Congress Cataloging-in-Publication Data

Names: Hristov, Jordan, editor.
Title: A closer look at the diffusion equation / Jordan Hristov, (editor),
  Professor of Chemical Engineering, Department of Chemical Engineering,
  University of Chemical Technology and Metallurgy, Sofia, Bulgaria.
Identifiers: LCCN 2020034029 (print) | LCCN 2020034030 (ebook) | ISBN
  9781536183306 (paperback) | ISBN 9781536184884 (adobe pdf)
Subjects: LCSH: Heat equation.
Classification: LCC QA377 .C56 2020 (print) | LCC QA377 (ebook) | DDC 515/.353--dc23
LC record available at https://lccn.loc.gov/2020034029
LC ebook record available at https://lccn.loc.gov/2020034030

Published by Nova Science Publishers, Inc. † *New York*

CONTENTS

Preface		vii
Chapter 1	A Numerical Approach to Solving Unsteady One-Dimensional Nonlinear Diffusion Equations *István Faragó, Stefan M. Filipov, Ana Avdzhieva and Gabriella Svantnerné Sebestyén*	1
Chapter 2	Diffusion in Hypersonic Flows *H. Berk Gür and Sinan Eyi*	27
Chapter 3	On the Nonlinear Diffusion with Exponential Concentration-Dependent Diffusivity: Integral-Balance Solutions and Analyzes *Jordan Hristov*	55
Chapter 4	Solutions for Fractional Reaction-Diffusion Equations *D. Marin, L. M. S. Guilherme, M. K. Lenzi, E. K. Lenzi and P. M. Ndiaye*	93
Chapter 5	Semi-Analytical Solution of Hristov Diffusion Equation with Source *Derya Avci and Beyza Billur Iskender Eroğlu*	117
Chapter 6	Non-Gaussian Diffusion Emergence in Superstatistics *Maike A. F. dos Santos*	133

| **Chapter 7** | Mean Square Displacement of the Fractional Diffusion Equation Described by Caputo Generalized Fractional Derivative
Ndolane Sene | **151** |

About the Editor   **175**

Index

  177

PREFACE

Diffusion is one of the principle transport mechanism of heat (energy) mass and momentum. Diffusion problems are classic challenge for engineers and scientists with consciously emerging interesting problems to be solved. Among the multifaceted applications of diffusion problems there are principle, say classical, problems that are always challenging to develop new solutions, among them: transient diffusion with non-linear diffusivities, wetting of porous media, diffusion approximation in high energy transfer, combustion, supersonic flows, etc.

This collection of studies arranged as a book was suggested by the publisher, and when the editor was invited there was an ambiguity what exactly to be included as topic. Then, we decided to invite authors who are appearing in journal publications and the responses as chapters arranged in this book clearly indicate what some of the modern trend in diffusion studies are.

Following these introductory notes, we may mention three chapters considering integer-order problem (local problems) (chapters 1, 2 and 3) and four studies devoted to time-fractional diffusion problems (Chapters 4, 5, 6 and 7) with different fractional operators. Thus, the collection is almost balanced from the point of view of problems presented and methods of solutions applied.

Farago et al. (Chapter 1) considered non-liner diffusion models with exponentially concentration depended diffusivities in both slow diffusion and fast diffusion modes. The numerical approach differs from the conventional ones when the equation is semi-discretized in space, which yields an initial-value (Cauchy) problem for a system of first order ODEs; In the method applied the first step is a discretization in time (by an implicit scheme), getting a sequence of two-point boundary value problems thus ensuring unconditionally stability of the solution approach. The second step addresses the nonlinear two-point boundary value problems where the finite difference method is applied successfully.

Diffusion in hypersonic flows is an interesting problem developed thoroughly by Gur and Evi in Chapter 2. Upon conditions of such flows air goes into chemical reaction due to high temperature. Hence, in addition to the Navier-Stokes Equations, chemical reaction equations should model the diffusion of the reaction chemical species. The study applies both the Fick's Law and the Stefan-Maxwell diffusion equation which differ in the formulations of the driving forces. Basically, in Fick's Law of Diffusion, the driving force is the species concentration differences. The results presented are evaluated by solving the Navier Stokes and finite-rate chemical reaction equations around the Apollo AS-202 Command module. The results reveal that the Stefan-Maxwell equation provides gives more detailed results, rather than the simplified Fick's law, since its formulation is more accurate and adequate to the circumstances in hypersonic flows.

Hristov in Chapter 3 applied integral-balance solutions to a diffusion problem with exponential concentration dependent diffusion model relevant to diffusion in polymers, soil writhing, chloride diffusion in concretes, etc. The solutions developed apply the method in two versions: Heat-balance integral (HBIM) and Double-integration method (DIM). The analysis of the problem solved and the consequent tests of the approximate solutions demonstrate the feasibility the integral balance method in solution of non-linear diffusion problems.

Marin et al. (Chapter 4) analyze solutions of different fractional reaction-diffusion equations, which can be related to irreversible or reversible processes. Such models can be obtained from the continuous-time random walk approach by considering suitable conditions connected to the reaction processes, e.g., of remotion of the walkers, in the case of irreversible processes, or on the probability density function to incorporate different fractional differential operators. In such cases exact solutions in terms of the Green function approach are obtained thus demonstrating how they are related to a rich variety of behavior related to anomalous diffusion.

Avci and Billur (Chapter 5) address a solution of one of the first type Hristov's diffusion equation with a source term. The model considered (with Atangana-Baleanu derivative of Riemann-Liouville type, ABR derivative), is related to the fading memory concept corresponding to the Boltzmann superposition principle. This recently conceived equation has a physically interpretable background unlike the directly fractionalized models. The developed semi-analytical solutions account effects of space and time dependent external sources. The approach applied reduces the model to ordinary differential equations successfully solved by the Fourier method. The resulting time-dependent ordinary fractional differential equation is solved approximately by the Diethelm's predictor-corrector algorithm.

Maike dos Santos (Chapter 6) addresses solutions of a non-Gaussian diffusion process that has a linear mean square displacement in time, also known as non-*Gaussian yet Brownian*. Such processes are atypical within anomalous diffusive processes. The non-Gaussian yet Brownian processes emerge in complex systems associated with biological, active, and soft matter systems. The problem developed in this chapter demonstrates how a superstatistical approach should be carried out when there are fluctuations in the diffusivities thus building-up overall distributions connected with non-Gaussian diffusion. Moreover, the chapter covers general distributions associated with three-parameter Mittag-Leffler distribution to diffusivities thus implying a new and broad class of overall distributions linked to non-Gaussian diffusion. Finally, the chapter develops superstatistics for a version of the Fokker-Planck equation to address system that considers fluctuations in diffusivity and a linear force in position.

The last Chapter 7 by N. Sene discusses Mean Square Displacement (MSD) of the fractional diffusion equation when the Caputo generalized fractional derivative is applied. The chapter presents an approximate method to determine the MSD and demonstrates it with practically relevant example.

After these brief outlines of the main results presented in the chapters collected, we may express out warm appreciation to all authors contributing the book. Moreover, last but not least, we would like to express our gratitude to the publishing office for their understanding and support about the problems emerging during the manuscript collection and editing. We believe that this book would be a good reference source to many researchers involved in various applications of diffusion problems and would motivate them to go further in this amazing and challenging area of research.

Prof. Dr. Jordan Hristov, PhD, DSc
Department of Chemical Engineering
University of Chemical Technology and Metallurgy
Sofia, Bulgaria
June 2020

In: A Closer Look at the Diffusion Equation
Editor: Jordan Hristov
ISBN: 978-1-53618-330-6
© 2020 Nova Science Publishers, Inc.

Chapter 1

A NUMERICAL APPROACH TO SOLVING UNSTEADY ONE-DIMENSIONAL NONLINEAR DIFFUSION EQUATIONS

István Faragó[1], *Stefan M. Filipov*[2,*],
Ana Avdzhieva[3] *and Gabriella Svantnerné Sebestyén*[1]
[1]Department of Differential Equations, Institute of Mathematics,
Budapest University of Technology and Economics,
& MTA-ELTE Research Group, Budapest, Hungary
[2]Department of Computer Science,
Faculty of Chemical and System Engineering,
University of Chemical Technology and Metallurgy, Sofia, Bulgaria
[3]Department of Numerical Methods,
Faculty of Mathematics and Informatics,
Sofia University St. Kliment Ohridski, Sofia, Bulgaria

Abstract

The diffusion problem has been widely investigated in the linear case, i.e., when the diffusion coefficient does not depend on the concentration. However, the problem with concentration-dependent diffusion coefficient, which results in nonlinear equation, is less investigated, but is also very important and is being studied actively. This work presents a numerical approach to solving the unsteady one-dimensional diffusion equation

---

*Corresponding Author's Email: sfilipov@uctm.edu.

with concentration-dependent diffusion coefficient. Unlike the traditional approach when the equation is semi-discretized in space, which yields an initial-value (Cauchy) problem for a system of first order ODEs, we first discretize the equation in time, getting a sequence of two-point boundary value problems. For the time-discretization we use an implicit scheme, which ensures that the method is unconditionally stable. To solve the obtained nonlinear two-point boundary value problems, the finite difference method is applied. For the solution of the arising nonlinear algebraic systems, the Newton iterative method is used. A MATLAB code that implements the proposed approach is presented. The approach is tested on various cases for the dependence of the diffusion coefficient on the concentration.

Keywords: nonlinear diffusion, concentration-dependent diffusion coefficient, partial differential equation, boundary value problem, finite difference method

1. INTRODUCTION

Diffusion [1]-[3] is the net movement of particles (e.g., atoms, ions, molecules) from a region where the concentration of the particles is higher to a region where the concentration is lower. Diffusion is present when there is a gradient in concentration. The concept of diffusion is widely used in many fields, including physics (particle diffusion), chemistry (diffusion of atoms and molecules), biology, sociology, economics, and finance (diffusion of people, ideas, and price values). The central idea of diffusion, however, is common to all of these: objects (e.g., particles, ideas, etc.) that are clustered together gradually, as time passes by, spread out to regions where the concentration of these objects is lower.

Each model of diffusion expresses the diffusion flux through concentrations, densities, and their derivatives. The diffusion coefficient (diffusivity, mass diffusivity) is a proportionality constant between the flux of a certain species and the gradient in the concentration of the species. Diffusivity is encountered by several equations e.g., Fick law describes diffusion of an admixture in a medium, Onsager's equation is used for multicomponent diffusion and thermo–diffusion. An approximate dependence of the diffusion coefficient on temperature in liquids can often be found using Stokes-Einstein equation, the dependence of the diffusion coefficient on temperature for gases can be expressed using Chapman-Enskog theory, etc. As a result, the mathematical models are some partial differential equations for the unknown function u (concentration), which depends on the space-variable $x \in \mathbb{R}^m$ (position) and the time-variable t (time). In this work, the one-dimensional case, i.e., $m = 1$, is considered. The obtained equations contain the diffusion coefficient, denoted by D. The diffusion coefficient may depend on position and time, but also on the (unknown) concentration. This means that, in general, $D = D(u, x, t)$. In the simpler case, when D does not depend on u, i.e., $D = D(x, t)$, the model is linear, otherwise it is nonlinear.

Nonlinear problems arise in many scientific and engineering fields, for example, in heat and mass transfer, fluid mechanics, thermo-elasticity, plasma physics, chemical physics, etc. Due to their great importance, nonlinear problems are subject to extensive research. For a number of cases analytical solutions to the governing equations have been found [4]-[8], but most often it is very difficult or impossible to find the exact analytical solution. Some approximate analytical methods have been proposed, such as Exp-function method [9]-[10], homotopy-perturbation method [11]-[14], variational iteration method (VIM) [15]-[17]. Many particular cases have been investigated/solved through different analytical techniques, e.g., [18]-[22] and the references therein. Nevertheless, the numerical methods for solving nonlinear problems remain a very important tool in investigating nonlinear phenomena [23]-[26]. In this paper we propose a new numerical approach for solving the unsteady, one–dimensional diffusion equation

$$\frac{\partial u}{\partial t} = \frac{\partial}{\partial x}\left(D\frac{\partial u}{\partial x}\right), \qquad (1)$$

where $u = u(x, t)$ denotes the (unknown) concentration of a certain species, x is the position, t is the time, and D is the diffusion coefficient. If D does not depend on the concentration u, then (1) a linear differential equation. Many numerical methods for solving the linear equation have been developed and investigated extensively. For example, the method of lines (MOL) [27]-[28], the finite element method [29]-[30], etc. Often, however, the diffusion coefficient D is concentration-dependent. Then, eqn. (1) is a nonlinear equation. The proposed in this work approach considers the case when D does not depend explicitly on position and time but depends on the concentration. This is an important case because it describes diffusion processes such as fast and slow diffusion [6], [16], [23], [31]-[33], diffusion in polymers, etc. Assuming $D = D(u)$ and performing the differentiation on the right hand side of (1) we get

$$\frac{\partial u}{\partial t} = \partial_u D(u) \left(\frac{\partial u}{\partial x}\right)^2 + D(u)\frac{\partial^2 u}{\partial x^2}. \tag{2}$$

Equation (2) will be solved on the spatial interval $[a, b]$ subject to the boundary conditions

$$u(a, t) = \alpha(t), \ u(b, t) = \beta(t), \ t > 0, \tag{3}$$

which give the concentration at the two ends as functions of time, and the initial condition

$$u(x, 0) = u_0(x), \ x \in [a, b], \tag{4}$$

which specifies the distribution of the concentration along the spatial interval at the beginning of the diffusion process.

2. IMPLICIT TIME DISCRETIZATION

An important and widely used numerical method for solving equation (2), subject to the boundary conditions (3) and the initial condition (4), is the method of lines (MOL). It was originally introduced in 1965 by Liskovetz [27] for partial differential equations of elliptic, parabolic, and hyperbolic type. Since then, it has been applied in many different situations and is still actively used today [28], [34]-[38]. The MOL first discretizes equation (2) in space (Figure 1), which yields an initial-value (Cauchy) problem for a system of first order ODEs. Then, the initial-value problem is solved by using some appropriate numerical method. If the explicit Euler method is used, then the combined method, when D is a constant, is stable only for $0 < \frac{D\tau}{h^2} \leq \frac{1}{2}$, where D is the diffusion coefficient, h is the discretization step in space, and τ is the discretization step in time. We propose an alternative approach for the solution of (2)-(4). Instead of discretizing equation (2) in space, we first discretize the equation in time (Figure 2). The time line $t \geq 0$ is divided into intervals of equal size by the mesh-points

$$t_n = n\tau, \ n = 0, 1, 2, ... \tag{5}$$

where τ is the step-size of the mesh. Then, equation (2) is discretized on the mesh (5) using an implicit scheme to approximate the time-derivative:

$$\frac{u_n - u_{n-1}}{\tau} = \partial_u D(u_n)\left(\frac{du_n}{dx}\right)^2 + D(u_n)\frac{d^2 u_n}{dx^2}, \tag{6}$$

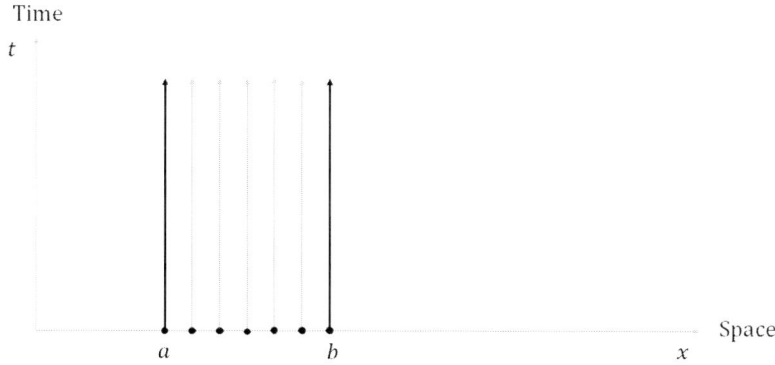

Figure 1. Space-discretization - Method of Lines (MOL). The values of u along the black solid lines (BCs) and at the black dots (IC) are given. The values of u along the blue lines should be found.

where $u_n = u_n(x)$ and $u_{n-1} = u_{n-1}(x)$ approximate the values of $u(x, t_n)$ and $u(x, t_{n-1})$, respectively. Equation (6) is a second order ordinary differential equation. It approximates the partial differential equation (2). The error is $O(\tau)$, hence, with respect to u_n, the discretization scheme is first-order accurate in time. The proposed discretization provides for unconditional stability of the method, unlike the explicit discretization scheme when the right-hand side of (2) is approximated by its value at $t = t_{n-1}$. This approach results in a method that is only conditionally stable.

We can write equation (6) in the form

$$\frac{d^2 u_n}{dx^2} = f(u_n, v_n; u_{n-1}), \qquad (7)$$

where $v_n = \frac{du_n}{dx}$ is the concentration gradient and f is the following nonlinear function:

$$f(u_n, v_n; u_{n-1}) = \frac{\phi(u_n, v_n; u_{n-1})}{D(u_n)}, \qquad (8)$$

$$\phi(u_n, v_n; u_{n-1}) = \frac{u_n - u_{n-1}}{\tau} - \partial_u D(u_n) v_n^2. \qquad (9)$$

Equation (7) and the boundary conditions $u_n(a) = \alpha(t_n)$, $u_n(b) = \beta(t_n)$ define a nonlinear two-point boundary value problem (TPBVP) for the unknown

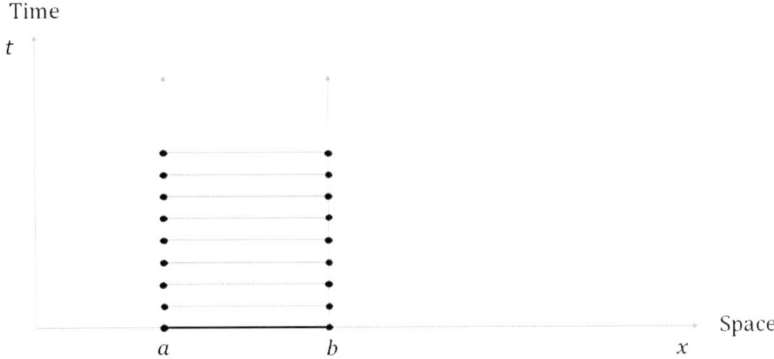

Figure 2. Time-discretization. The values of u along the black solid line (IC) and at the black dots (BCs) are given. The values of u along the blue lines should be found.

function u_n. If the function u_{n-1} is known (given), the problem can be solved by using some numerical technique for nonlinear problems. Thus, starting from the initial concentration distribution u_0 we can solve successively (7), subject to the given boundary conditions, getting the concentration distributions u_n at time t_n, $n = 1, 2, ...$

3. SOLVING THE NONLINEAR TPBVPS

To solve the obtained nonlinear two-point boundary value problems, an appropriate numerical method and iterative procedure should be chosen and implemented. We propose the finite difference method (FDM) with the Newton method. This section presents the implementation of these methods and discusses their benefits in the particular situation.

3.1. Partial Derivatives for the Newton Method

A good choice of an iterative method is the Newton method (Newton-Raphson method) [39] because it is reliable and fast. Its convergence is quadratic. Sometimes, it may suffer convergence problems, especially around extrema and inflection points [40], but these situations are rare and when they occur one can easily change the obtained general Newton iteration formula to Picard iteration, which may (sometimes) overcome the problem. How it can be done is discussed in the next section. The implementation of the Newton method requires the partial derivatives of the function $f(u_n, v_n; u_{n-1})$ with respect to u_n and v_n. Introducing the notation $f_n = f(u_n, v_n; u_{n-1})$, $\phi_n = \phi(u_n, v_n; u_{n-1})$ and denoting the partial derivatives of f_n by $q_n = q(u_n, v_n; u_{n-1})$, $p_n = p(u_n, v_n)$ we get:

$$q_n = \frac{\partial f_n}{\partial u_n} = \frac{1}{D(u_n)}\left(-f_n \partial_u D(u_n) + \frac{\partial \phi_n}{\partial u_n}\right), \qquad (10)$$

$$p_n = \frac{\partial f_n}{\partial v_n} = \frac{1}{D(u_n)} \frac{\partial \phi_n}{\partial v_n}, \qquad (11)$$

where

$$\frac{\partial \phi_n}{\partial u_n} = \frac{1}{\tau} - \partial_{uu}^2 D(u_n) v_n^2, \qquad (12)$$

$$\frac{\partial \phi_n}{\partial v_n} = -2\partial_u D(u_n) v_n. \qquad (13)$$

These formulas will be necessary later on in section 3.3 when calculating the elements of the Jacobian matrix for the Newton finite difference (relaxation) method.

3.2. Applying the Finite Difference Method

Choosing the right numerical method for the solution of equation (7), subject to boundary conditions $u_n(a) = \alpha(t_n)$, $u_n(b) = \beta(t_n)$, is very important. Although the shooting method is fast and easy to implement [40]-[42], it is not appropriate for this particular case because refining the time-mesh, i.e., decreasing the time step τ, causes the term $\frac{1}{\tau}$ in (9) to grow unboundedly, which effects the IVP solutions [42]. It turns out that the finite difference method (FDM) [39], [43]- [44] is a much better choice. Hence, we have adopted the FDM. Let us divide the interval $[a, b]$ by N equally separated mesh-points

$$x_i = a + (i-1)h, \quad i = 1, \ldots, N; \quad h = \frac{b-a}{N-1}. \qquad (14)$$

The points (14) define a uniform mesh on the interval $[a, b]$. Equation (7) is discretizes on the mesh (14) by using the FDM with the central difference approximation for the second derivative:

$$\frac{u_{n,i+1} - 2u_{n,i} + u_{n,i-1}}{h^2} = f(u_{n,i}, v_{n,i}; u_{n-1,i}), \quad i = 2, 3, \ldots, N-1, \qquad (15)$$

where
$$v_{n,i} = \frac{u_{n,i+1} - u_{n,i-1}}{2h}. \tag{16}$$

In (15) and (16) $u_{n,i}$, $v_{n,i}$, and $u_{n-1,i}$ denote approximations to the exact values $u_n(x_i)$, $v_n(x_i)$, and $u_{n-1}(x_i)$. Correspondingly, everywhere in equations (8)-(13), we set $x = x_i$ and then replace the exact values at the mesh-points with their approximations $u_{n,i}$, $v_{n,i}$, and $u_{n-1,i}$. Equation (15) holds for the inner mesh-points and approximates (7) with error $O(h^2)$, i.e., it is second-order accurate in space. At the boundaries the boundary conditions should be applied:

$$u_{n,1} = \alpha(t_n), \ u_{n,N} = \beta(t_n). \tag{17}$$

3.3. Solving the Nonlinear System by Newton Method

Equations (15), together with the boundary conditions (17), constitute a nonlinear system of N algebraic equations for the N unknowns $u_{n,i}$, $i = 1, 2, \ldots, N$. To solve this nonlinear system, the Newton iterative method will be applied. Introducing the column-vector $\mathbf{G}_n = [G_{n,1}, G_{n,2}, \ldots, G_{n,N}]^T$ with components

$$G_{n,1} = u_{n,1} - \alpha(t_n), G_{n,N} = u_{n,N} - \beta(t_n), \tag{18}$$

$$G_{n,i} = u_{n,i+1} - 2u_{n,i} + u_{n,i-1} - h^2 f_{n,i}, \tag{19}$$

$$f_{n,i} = f(u_{n,i}, v_{n,i}; u_{n-1,i}), i = 2, 3, \ldots, N-1, \tag{20}$$

equations (15) and the boundary conditions (17) are written as one equation:

$$\mathbf{G}_n(\mathbf{u}_n) = 0, \tag{21}$$

where
$$\mathbf{u}_n = [u_{n,1}, u_{n,2}, \ldots, u_{n,N}]^T. \tag{22}$$

The Newton iteration for the system (21) is

$$\mathbf{u}_n^{(k+1)} = \mathbf{u}_n^{(k)} - (\mathbf{L}_n^{(k)})^{(-1)} \mathbf{G}_n(\mathbf{u}_n^{(k)}), k = 0, 1, 2, \ldots \tag{23}$$

where $\mathbf{L}_n^{(k)}$ is the Jacobian of $\mathbf{G}_n$ with respect to $\mathbf{u}_n$ evaluated at $\mathbf{u}_n^{(k)}$:

$$\mathbf{L}_n^{(k)} = \frac{\partial \mathbf{G}_n}{\partial \mathbf{u}_n}(\mathbf{u}_n^{(k)}). \tag{24}$$

The elements of the Jacobian $\mathbf{L}_n^{(k)}$ are:

$$L_{n(1,1)}^{(k)} = 1, L_{n(N,N)}^{(k)} = 1, \qquad (25)$$

$$L_{n(i,i-1)}^{(k)} = 1 + \frac{1}{2}hp_{n,i}^{(k)}, L_{n(i,i)}^{(k)} = -2 - h^2 q_{n,i}^{(k)}, L_{n(i,i+1)}^{(k)} = 1 - \frac{1}{2}hp_{n,i}^{(k)}, \quad (26)$$

$$q_{n,i}^{(k)} = q(u_{n,i}^{(k)}, v_{n,i}^{(k)}; u_{n-1,i}), p_{n,i}^{(k)} = p(u_{n,i}^{(k)}, v_{n,i}^{(k)}), i = 2, 3, \ldots, N-1. \quad (27)$$

The derivation of (26) is given in the Appendix. To calculate $q_{n,i}^{(k)}$ and $p_{n,i}^{(k)}$ formulas (10)-(11) can be used. Iteration (23) is a one-step (two-level) iteration. Starting from some initial guess $\mathbf{u}_n^{(0)}$, we can find each next approximation $\mathbf{u}_n^{(k+1)}$, $k = 0, 1, 2, \ldots$ using (23). If the sequence is convergent, then the limiting vector $\mathbf{u}_n = \lim_{k\to\infty}(\mathbf{u}_n^{(k+1)})$ is a solution to the nonlinear system (21). If, for some reason, the Newton iteration turns out to be divergent, then one can try the Picard iteration, which is a fixed-point iteration. The Picard iteration is given by (23) with matrix elements (25), (26), where $q_{n,i}^{(k)} = 0$ and $p_{n,i}^{(k)} = 0$.

In practice, the iteration process is carried out until some stopping criteria of the form

$$\|\mathbf{u}_n^{(k+1)} - \mathbf{u}_n^{(k)}\| < \varepsilon \qquad (28)$$

is satisfied. In the numerical examples provided in the next section, for a norm in (28) we have used the discretized L^2-norm. The vector $\mathbf{u}_n^{(k+1)}$ is taken as approximate solution to (21). As an initial guess $\mathbf{u}_n^{(0)}$, we use the solution $\mathbf{u}_{n-1}$ found at the previous time-step.

4. NUMERICAL COMPUTER EXPERIMENTS

In this section the proposed approach is tested on several cases. In Example 1, equation (1) is solved, with given initial condition and boundary conditions, for $D = u^{-2}$ (fast diffusion). The result is compared with the exact (analytical) solution and the convergence of the method is investigated as the time-mesh is refined. In Example 2, $D = 1$ (linear diffusion) is compared with $D = u^m$, $m = 1, 2$ (slow diffusion). Finally, in Example 3, the case $D = D_0 e^{cu}$ is investigated for both positive and negative values of the constant c.

Example 1. Fast diffusion $D = u^{-2}$.

Any diffusion with diffusion coefficient $D = u^m, m < 0$ is called fast diffusion [6], [7], [16], [23], [31]-[33]. In this example we consider equation (1) with diffusion coefficient

$$D = \frac{1}{u^2}, \qquad (29)$$

initial condition

$$u(x, 0) = \frac{1}{\sqrt{x^2 + 1}}, \qquad (30)$$

and the following time-dependent boundary conditions on the space-domain $[-10, 10]$:

$$u(-10, t) = u(10, t) = \frac{1}{\sqrt{100 + e^{2t}}}. \qquad (31)$$

The exact (analytical) solution to the problem (29)-(31) is

$$u(x, t) = \frac{1}{\sqrt{x^2 + e^{2t}}}. \qquad (32)$$

The differential equation (1) with diffusion coefficient (29) and initial condition (30), on the unbounded space-domain $(-\infty, +\infty)$, has the same analytical solution (32) [23].

The problem (1), (29)-(31) is solved by the proposed in this work numerical approach on the time interval $[0, T]$ with time-step

$$\tau = \frac{T}{M - 1}, \qquad (33)$$

where T is the final time and M is the number of mesh-point on the time-interval. First, the problem is solved for $T = 5$, $M = 31$, and number of space mesh-points $N = 41$. A $3D$ plot of the concentration $u(x, t)$ as a function of x and t (time) is shown in Figure 3.

Side-views (2D projections) of the plot in Figure 3 are presented in Figure 4 and Figure 5. The concentration u as function of x is shown (Figure 4) for times, starting from top to bottom, $t_n = n\tau, n = 0, 1, \ldots, 30$, where $\tau = 1/6$. The concentration u as a function of time is shown (Figure 5) for positions, starting from top to bottom, $x = 0, \pm h, \pm 2h, \ldots, \pm 20h$, where $h = 1/2$. To investigate the convergence properties of the method, the same problem, namely (1) with diffusion coefficient (29), initial condition (30) and boundary conditions

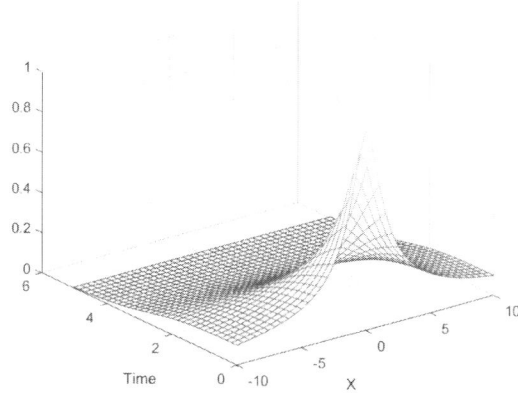

Figure 3. The concentration $u(x,t)$ for $D = u^{-2}$.

(31), is solved for a varying number of time-points, while the number of space-points is kept fixed. The deviation of the numerical solution from the exact solution at the final time $T = 2$ is investigated. The number of space-points is taken to be $N = 2001$, which corresponds to $h = 0.01$. For the number of time-points we have used $M = 2^l.20 + 1, l = 0, 1, 2, \ldots, 5$, which correspond to time-steps $\tau = 10^{-1}.2^{-l}$. When solving the a TPBVP at each time-level the iteration process is stopped when the difference between two successive approximations (28) becomes less than $\epsilon = 10^{-4}$. Since the method is second-order accurate in space, i.e., the error is $O(h^2)$, the chosen number of space-points is large enough and guarantees that the introduced error due to the finite space-discretization is less than ϵ. At the final time $T = 2$, the numerical solution is compared with the exact solution. The error, for a given time-step τ, is defined as the maximum norm of the difference between the exact and the numerical solution

$$\epsilon_\tau = \|u^{exact}(x, T) - u_\tau^{num}(x, T)\|_\infty. \tag{34}$$

The error ϵ_τ as a function of the time-step τ is shown in Table 1. The results indicate that decreasing the time-step twice reduces the error twice, hence the error is $O(\tau)$. This means that the proposed numerical method is, as expected, first order accurate in time. How the numerical solution at time $T = 2$ approaches the exact solution at that time as the time-step decreases is shown in Figure 6. A MATLAB code for the solution of (1) with diffusion coefficient

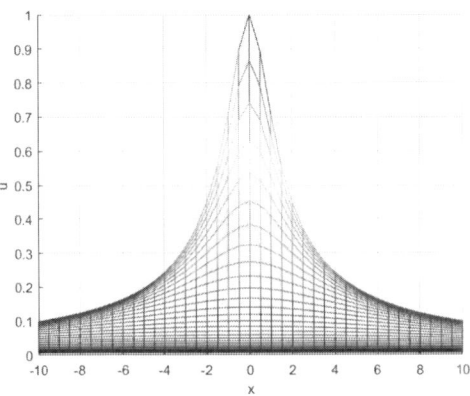

Figure 4. Profiles u vs x for $D = u^{-2}$.

Table 1. The error ϵ_τ (34) as a function of the time-step τ

M	τ	ϵ_τ
21	1/10	0.0030
41	1/20	0.0015
81	1/40	7.55×10^{-4}
161	1/80	3.79×10^{-4}
321	1/160	1.90×10^{-4}
641	1/320	9.54×10^{-5}

(29), initial condition (30), and boundary conditions (31) is given in section 5.

Example 2. Linear diffusion $D = 1$ and slow diffusion $D = u^m$, $m = 1, 2$.

Any diffusion for which $D = u^m$, $m > 0$ is called slow diffusion [16], [23], [31]- [33]. This example compares the linear diffusion $D = 1$ with the slow diffusions $D = u$ and $D = u^2$. Using the proposed approach, equation (1) is solved numerically on the spatial domain $x \in [0, 1]$ for initial condition

$$u(x, 0) = 1 \tag{35}$$

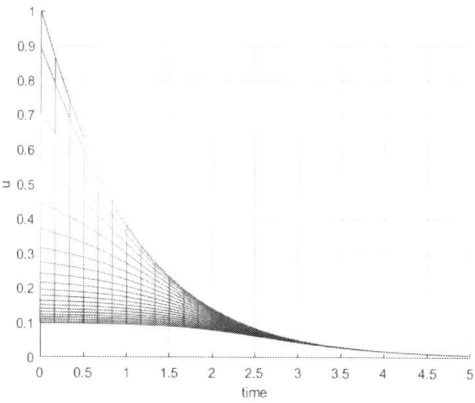

Figure 5. Profiles u vs t for $D = u^{-2}$.

and boundary conditions

$$u(0,t) = 0, u(1,t) = 0. \tag{36}$$

The finial time is chosen to be $T = 1$. The number of time mesh-points is $M = 31$ and the number of space mesh-points is $N = 41$. The results are shown in Figure 7. The side-views (profiles) u versus x are shown in Figure 8.

Example 3. Diffusion with diffusion coefficient $D = D_0 e^{cu}$.

This example investigates equation (1) with exponentially depending on u diffuson coefficient:

$$D = D_0 e^{cu} \tag{37}$$

where $D_0 = 0.1$. The following values of the constant c are considered:

$$c = -1, -0.5, 0, 0.5, 1. \tag{38}$$

Using the proposed numerical approach, equation (1) with diffusion coefficient (37) and values of c (38) is solved on the spatial domain $x \in [0, 1]$ for initial condition

$$u(x, 0) = 0 \tag{39}$$

and boundary conditions

$$u(0, t) = 1, u(1, t) = 0. \tag{40}$$

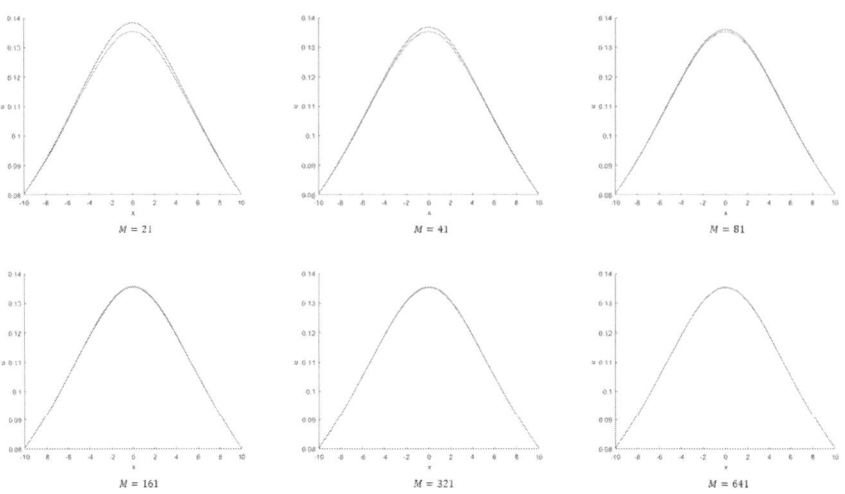

Figure 6. The exact solution at time $T = 2$ (red) and the numerical (blue) for increasing number of time mesh-points M.

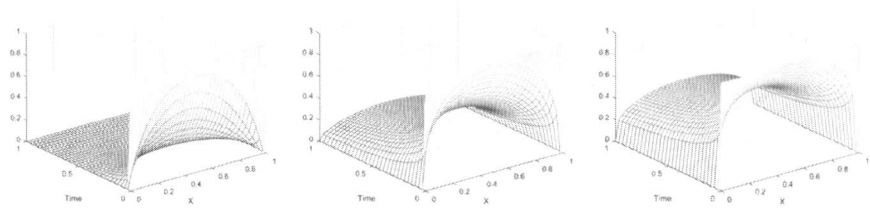

Figure 7. The concentration $u(x, t)$ for $D = 1$, $D = u$, and $D = u^2$.

The final time used in the experiments is $T = 15$. The number of time mesh-points is $M = 31$ and the number of space mesh-points is $N = 41$. Results for $c = -1, -0.5, 0$ are shown in Figure 9. The corresponding side-views (profiles) u versus x are shown in Figure 10. Results for $c = 0, 0.5, 1$ are shown in Figure 11. The corresponding side-views (profiles) u versus x are shown in Figure 12. The final concentration distribution reached in the computer experiments for cases $c = 0$, $c = 0.5$, and $c = 1$ (Figure 12) is practically the stationary distribution. For cases $c = -1$ and $c = -0.5$, a little bit more time is need. As can be seen from the figures, the stationary distribution for positive

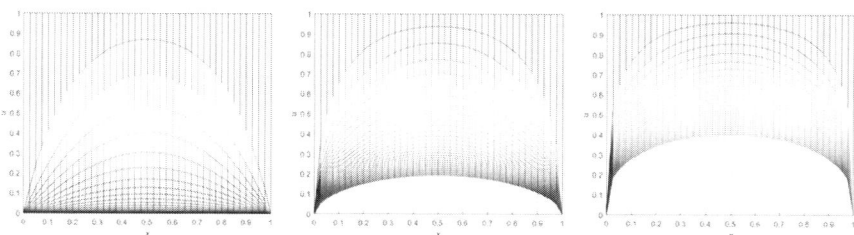

Figure 8. Profiles u vs x for $D = 1$, $D = u$, and $D = u^2$.

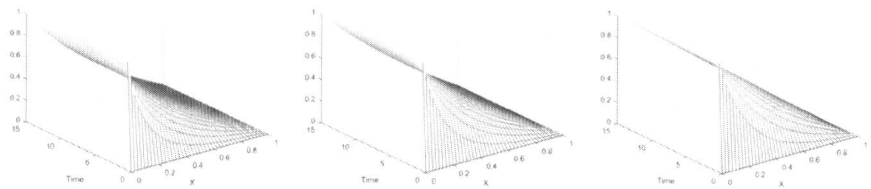

Figure 9. The concentration $u(x,t)$ for $D = 0.1e^{cu}$ and $c = -1, c = -0.5, c = 0$.

values of c is a concave function of x, for negative values of c the distribution is a convex function, while for $c = 0$ (linear diffusion) it is a straight line. In addition to these experiments we have performed an experiment for diffusion coefficient (37) with $c = 1$, final time $T = 1$, time discretization $M = 91$, and space discretization $N = 91$. The concentration profiles are shown in Figure 13. The final distribution reached in the experiment exhibits a typical shape characteristic for diffusion processes with diffusion coefficient which depends exponentially on the concentration.

5. MATLAB CODE

In this section a MATLAB code that implements the proposed numerical approach is presented. The code was used for solving Example 1 in the previous section. It can easily be modified for other cases. The mathematical notations used in this paper and the corresponding MATLAB variables used in the code are listed in Table 2. To copy the code in MATLAB, please open

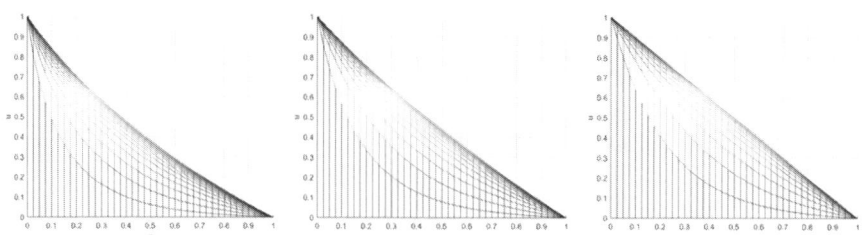

Figure 10. Profiles u vs x for $D = 0.1e^{cu}$ and $c = -1, c = -0.5, c = 0$.

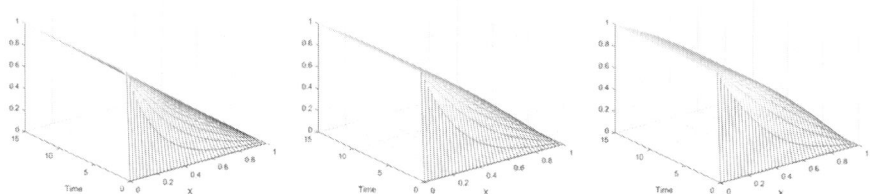

Figure 11. The concentration $u(x,t)$ for $D = 0.1e^{cu}$ and $c = 0, c = 0.5, c = 1$.

the file with Acrobat Reader. After pasting the code in the MATLAB editor window you may need to recover the backslash operator sign by hand.

```
function main
    M=51; N=81;
    tEnd=5; tau=tEnd/(M-1); A=1/tau;
    a=-10; b=10; h=(b-a)/(N-1);
    x=zeros(N,1);
    u0=zeros(N,1);
    for i=1:N
        x(i)=a+(i-1)*h;
        u0(i)=1/sqrt(x(i)*x(i)+1);
    end
    t=zeros(M,1);
    alpha=zeros(M,1); beta=zeros(M,1);
    for n=1:M
        t(n)=(n-1)*tau;
```

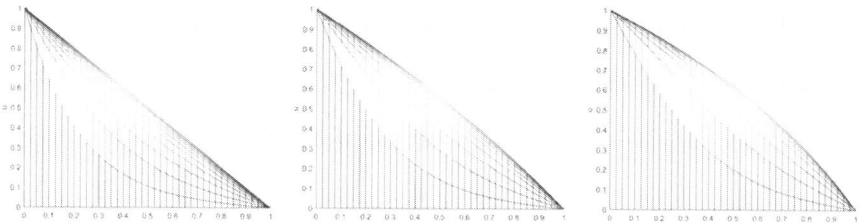

Figure 12. Profiles u vs x for $D = 0.1e^{cu}$ and $c = 0, c = 0.5, c = 1$.

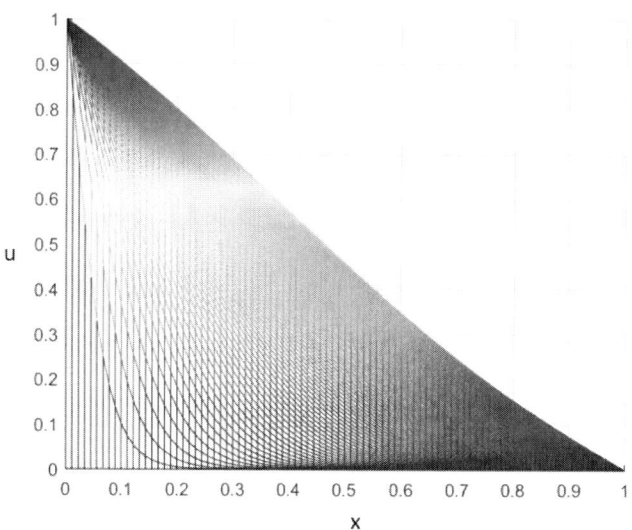

Figure 13. Concentration profiles u vs x for $D = 0.1e^u$ and final time $T = 1$.

```
   alpha(n)=1/sqrt(a*a+exp(2*t(n)));
   beta(n)=1/sqrt(b*b+exp(2*t(n)));
 end
 u1=zeros(N,1); u=zeros(N,1);
 uNext=zeros(N,1); G=zeros(N,1);
 L=zeros(N,N); L(1,1)=1; L(N,N)=1;
 U=zeros(N,M); U(:,1)=u0;
 for n=2:M
   u=U(:,n-1); u1=U(:,n-1);
   eps=1;
   while(eps>0.0001)
     G(1)=u(1)-alpha(n);
     G(N)=u(N)-beta(n);
      for i=2:N-1
        D=u(i)^(-2);
        dD=(-2)*u(i)^(-3);
        d2D=6*u(i)^(-4);
        v=(u(i+1)-u(i-1))/(2*h);
        phi=A*(u(i)-u1(i))-dD*v*v;
        f=phi/D;
        q=(-f*dD+A-d2D*v*v)/D;
        p=-2*dD*v/D;
        G(i)=u(i+1)-2*u(i)+u(i-1)-h*h*f;
        L(i,i-1)=1+0.5*h*p;
        L(i,i)=-2-h*h*q;
        L(i,i+1)=1-0.5*h*p;
      end
     uNext=u-L\G;
     eps=sqrt(h*(uNext-u)'*(uNext-u));
      u=uNext;
    end
     U(:,n)=u;
  end
  mesh(x,t,U');
end
```

Table 2. Mathematical notations and corresponding MATLAB variables used in the code

Mathematical notation	MATLAB	Mathematical notation	MATLAB
τ	tau	$\mathbf{u}_{n-1}$	u1
T	tEnd	$u_{n,i}$	U(i,n+1)
N	N	$\mathbf{G}_n(\mathbf{u}_n^{(k)})$	G
M	M	$\mathbf{L}_n^{(k)}$	L
a	a	$\|\mathbf{u}_n^{(k+1)} - \mathbf{u}_n^{(k)}\|$	eps
b	b	$u_{n,i}^{(k)}$	u(i)
h	h	$D(u_{n,i}^{(k)})$	D
t_n	t(n+1)	$\partial_u D(u_{n,i}^{(k)})$	dD
α_n	alpha(n+1)	$\partial_{uu}^2 D(u_{n,i}^{(k)})$	d2D
β_n	beta(n+1)	$v_{n,i}^{(k)}$	v
x_i	x(i)	$\phi(u_{n,i}^{(k)}, v_{n,i}^{(k)}; u_{n-1,i})$	phi
$\mathbf{u}_0$	u0	$f(u_{n,i}^{(k)}, v_{n,i}^{(k)}; u_{n-1,i})$	f
$\mathbf{u}_n^{(k)}$	u	$q_{n,i}^{(k)}$	q
$\mathbf{u}_n^{(k+1)}$	uNext	$p_{n,i}^{(k)}$	p

APPENDIX

In the calculations below, for simplicity, we have dropped some of the indexes. Let

$$G_i = u_{i+1} - 2u_i + u_{i-1} - h^2 f(u, v), \quad i = 2, 3, \ldots, N-1,$$

where

$$u = u_i, \quad v = \frac{u_{i+1} - u_{i-1}}{2h},$$

and let $q = \partial_u f$ and $p = \partial_v f$. Then, the nonzero elements of the Jacobian matrix $\mathbf{L}$ for rows $2, 3, \ldots, N-1$ are:

$$L_{i,i-1} = \frac{\partial G_i}{\partial u_{i-1}} = 1 - h^2 \Big(\frac{\partial f}{\partial u}\frac{\partial u}{\partial u_{i-1}} + \frac{\partial f}{\partial v}\frac{\partial v}{\partial u_{i-1}}\Big) = 1 - h^2 (q.0 - p.\frac{1}{2h}) = 1 + \frac{hp}{2},$$

$$L_{i,i} = \frac{\partial G_i}{\partial u_i} = -2 - h^2 \Big(\frac{\partial f}{\partial u}\frac{\partial u}{\partial u_i} + \frac{\partial f}{\partial v}\frac{\partial v}{\partial u_i}\Big) = -2 - h^2 (q.1 + p.0) = -2 - h^2 q,$$

$$L_{i,i+1} = \frac{\partial G_i}{\partial u_{i+1}} = 1 - h^2 \Big(\frac{\partial f}{\partial u}\frac{\partial u}{\partial u_{i+1}} + \frac{\partial f}{\partial v}\frac{\partial v}{\partial u_{i+1}}\Big) = 1 - h^2(q.0 + p.\frac{1}{2h}) = 1 - \frac{hp}{2}.$$

Rows 1 and N are for the boundary conditions. If we want to change the boundary conditions from Dirichlet to something else (e.g., Neumann, general linear, nonlocal, integral condition, etc.) we just need to change these two rows [39].

CONCLUSION

This work presented a new numerical approach to solving the unsteady one-dimensional diffusion equation with concentration-dependent diffusion coefficient. The approach relies on implicit time-discretization, which provides for unconditional stability of the method. For the solution of the arising nonlinear two-point boundary value problems, the finite difference method was applied. The nonlinear systems of algebraic equations were solved by the Newton method. In case the Newton method turns out to be divergent for a particular nonlinearity, it can easily be replaced by the Picard iterative method. The numerical approach allows easy incorporation of different types of boundary conditions including general linear, nonlocal, and integral conditions. The incorporation of nonlinear boundary conditions is also possible. A MATLAB code implementing the proposed method was presented. The code can easily be modified to implement different iterative methods and to incorporate different types of boundary conditions. Several examples were solved demonstrating the efficiency and reliability of the proposed approach.

Acknowledgments

This research was carried out in the ELTE Institutional Excellence Program (1783-3/2018/FEKUTSRAT) supported by the Hungarian Ministry of Human Capacities, and supported by the Hungarian Scientific Research Fund OTKA, SNN125119. The research was also supported by the European Union, and co-financed by the European Social Fund (EFOP-3.6.3-VEKOP-16-2017-00002). The work of Stefan Filipov and Ana Avdzhieva was supported by the Bulgarian Ministry of Education and Science under the National Research Programme "Young scientists and postdoctoral students".

REFERENCES

[1] Mehrer, H., *Diffusion in solids: fundamentals, methods, materials, diffusion-controlled processes*, Springer Science & Business Media, Vol. 155, 2007. https://www.springer.com/gp/book/9783540714866.

[2] Bird, R. B., Stewart, S. E. and Lightfoot, E. N., *Transport Phenomena*, Revised 2nd Edition, John Wiley & Sons, Inc.; 2nd edition, 2006.

[3] https://en.wikipedia.org/wiki/Diffusion.

[4] Ames, W.F., *Nonlinear Partial Differential Equations*, Academic Press, New York, 1972.

[5] Wazwaz, A.M., Exact solutions to nonlinear diffusion equations by the decomposition method, *Appl. Math. Comput.* **123**, 1, 109-122 (2001). https://doi.org/10.1016/S0096-3003(00)00064-3.

[6] Wazwaz, A.M., Several new exact solutions for a fast diffusion equation by the differential constraints of the linear determining equations, *Appl. Math. Comput.* **145**, 525-540 (2003). https://doi.org/10.1016/S0096-3003(02)00512-X.

[7] Polyanin, A.D. and Zaitsev, V.F., *Handbook of Nonlinear Partial Differential Equations*, Second Edition, Chapman & Hall/CRC Press, Boca Raton-London-New York, 2012.

[8] Kosov, A.A. and Semenov, É.I., Exact Solutions of the Nonlinear Diffusion Equation, *Sib Math J* **60**, 93 (2019). https://doi.org/10.1134/S0037446619010117.

[9] He, J.H. and Wu, X.-H., Exp-function method for nonlinear wave equations, *Chaos, Solitons and Fractals* **30**, 3, 700-708 (2006). https://doi.org/10.1016/j.chaos.2006.03.020.

[10] He, J.H. and Abdou, M.A., New periodic solutions for nonlinear evolution equations using Exp-function method, *Chaos, Solitons and Fractals* **34**, 5, 1421–1429 (2007). https://doi.org/10.1016/j.chaos.2006.05.072.

[11] He, J.H., A coupling method of a homotopy technique and a perturbation technique for non-linear problems, *International Journal of Non-Linear Mechanics* **35**, 1, 37-43 (2000). https://doi.org/10.1016/S0020-7462(98)00085-7.

[12] He, J.H., Homotopy perturbation method: A new nonlinear analytical technique, *Applied Mathematics and Computation* **135**, 1, 73-79 (2003). https://doi.org/10.1016/S0096-3003(01)00312-5.

[13] He, J.H., New interpretation of homotopy perturbation method, *International Journal of Modern Physics B* **20**, 18, 2561-2568 (2006). https://doi.org/10.1142/S0217979206034819.

[14] He, J.H., Homotopy perturbation method for solving boundary value problems, *Physics Letters A* **350**, 1-2, 87-88 (2006). https://doi.org/10.1016/j.physleta.2005.10.005.

[15] He, J.H., Variational iteration method - a kind of non–linear analytical technique: Some examples, *International Journal of Non-Linear Mechanics* **34**, 699-708 (1999). https://doi.org/10.1016/S0020-7462(98)00048-1.

[16] Sadighi, A. and Ganji, D.D., Exact solutions of nonlinear diffusion equations by variational iteration method, *Computers and Mathematics with Applications* **54**, 1112–1121 (2007). https://doi.org/10.1016/j.camwa.2006.12.077.

[17] Wazwaz, A.M., The variational iteration method: a powerful scheme for handling linear and nonlinear diffusion equations, *Computers & Mathematics with Applications* **54**, 933-939 (2007). https://doi.org/10.1016/j.camwa.2006.12.039.

[18] Hristov, J., An approximate analytical (integral-balance) solution to a nonlinear heat diffusion equation, *Thermal Science* **19**, 2, 723–733 (2015). DOI: 10.2298/TSCI140326074H http://www.doiserbia.nb.rs/img/doi/0354-9836/2015/0354-98361400074H.pdf.

[19] Hristov, J., Integral solutions to transient nonlinear heat (mass) diffusion with a power-law diffusivity: a semi-infinite medium with fixed boundary conditions, *Heat Mass Transfer* **52**, 3, 635-655 (2016). https://doi.org/10.1007/s00231-015-1579-2.

[20] Fabre, A. and Hristov, J., On the integral-balance approach to the transient heat conduction with linearly temperature-dependent thermal diffusivity, *Heat and Mass Transfer* **53**, 1, 177-204 (2017). https://doi.org/10.1007/s00231-016-1806-5.

[21] Fabre, A., Hristov, J. and Bennacer, R., Transient Heat Conduction in Materials with Linear Power–Law Temperature–Dependent Thermal Conductivity: Integral–Balance Approach, *Fluid Dynamics and Materials Processing* **12**, 1, 69–85 (2016). doi:10.3970/fdmp.2016.012.069 http://www.techscience.com/fdmp/v12n2/24614.

[22] Hristov, J., A new closed-form approximate solution to diffusion with quadratic Fujitas non-linearity: the case of diffusion controlled sorption kinetics relevant to rectangular adsorption isotherms, *Heat and Mass Transfer* **55**, 2, 261-279 (2019). https://doi.org/10.1007/s00231-018-2408-1.

[23] Grarslan, G., Numerical modelling of linear and nonlinear diffusion equations by compact finite difference method, *Applied Mathematics and Computation* **216**, 8, 2472–2478 (2010). https://doi.org/10.1016/j.amc.2010.03.093.

[24] Grarslan G. and Sari G., Numerical solutions of linear and nonlinear diffusion equations by a differential quadrature method (DQM), *Commun. Numer. Meth. En.* (2009). doi:10.1002/cnm.1292. https://doi.org/10.1002/cnm.1292.

[25] Dhar, D.K. and Das, S., Numerical solution of the nonlinear diffusion equation by using non–standard/standard finite difference and Fibonacci collocation methods, *Phys. J. Plus* **134**, 608 (2019). https://doi.org/10.1140/epjp/i2019-12953-x.

[26] Jannelli, A., Ruggieri, M. and Speciale, M. P., Numerical solutions of space–fractional advection-diffusion equations with nonlinear source term, *Applied Numerical Mathematics* (2020). https://doi.org/10.1016/j.apnum.2020.01.016.

[27] Liskovets, O.A., The Method of Lines, *J. Diff. Eqs.* **1**, 1308 (1965). http://www.mathnet.ru/links/bc14dc670dda8ce9ebf0c431f2287010/de9396.pdf.

[28] Zafarullah, A., Application of the Method of Lines to Parabolic Partial Differential Equations With Error Estimates, *Journal of the ACM* **17**, 2, 294–302 (1970). https://dl.acm.org/citation.cfm?id=321583.

[29] Vidar, T., *Galerkin Finite Element Methods for Parabolic Problems*, Springer-Verlag, 2006. DOI: 10.1007/3-540-33122-0. https://www.springer.com/gp/book/9783540331216.

[30] Larson, M.G. and Bengzon F., *The Finite Element Method: Theory, Implementation, and Applications*, Springer-Verlag, 2013. DOI: 10.1007/978-3-642-33287-6. https://www.springer.com/gp/book/9783642332869.

[31] Saeid, E.A., The nonclassical solution of the inhomogeneous non-linear diffusion equation, *Applied Mathematics and Computation* **98**, 103-108 (1999). https://doi.org/10.1016/S0096-3003(97)10158-8.

[32] Saeid, E.A. and Hussein, M.M., New classes of similarity solutions of the inhomogeneous nonlinear diffusion equations, *Journal of Physics A* **27**, 4867-4874 (1994). https://iopscience.iop.org/article/10.1088/0305-4470/27/14/015.

[33] Dresner, L., *Similarity Solutions of Nonlinear Partial Differential Equations*, Pitman, New York, 1983.

[34] Schiesser, W., *The Numerical Method of Lines*, Integration of Partial Differential Equations, 1991. https://www.elsevier.com/books/the-numerical-method-of-lines/schiesser/978-0-12-624130-3.

[35] Sommeijer, B., Shampine, L. and Verwer, J., An explicit solver for parabolic PDEs, *Journal of Computational and Applied Mathematics* **88**, Issue 2, 315–326 (1998). https://www.sciencedirect.com/science/article/pii/S0377042797002197.

[36] Marucho, M.D. and Campo, A., Suitability of the Method of Lines for rendering analytic/numeric solutions of the unsteady heat conduction equation in a large plane wall with asymmetric convective boundary conditions, *International Journal of Heat and Mass Transfer* **99**, 201-208 (2016). https://doi.org/10.1016/j.ijheatmasstransfer.2016.03.118.

[37] Kazem, S. and Dehgham, M., Application of finite difference method of lines on the heat equation, *Numerical Methods for Partial Differential Equations* **34**, 2, 626–660 (2017). https://onlinelibrary.wiley.com/doi/abs/10.1002/num.22218.

[38] Ramos, J., A conservative, piecewise-analytical, transversal method of lines for reaction-diffusion equations, *International Journal of Numerical Methods for Heat & Fluid Flow* **29**, 4093–4129 (2019). https://doi.org/10.1108/HFF-01-2019-0025.

[39] Filipov, S. M., Gospodinov, I. D. and Faragó, I., Replacing the finite difference methods for nonlinear two-point boundary value problems by successive application of the linear shooting method, *Journal of Computational and Applied Mathematics* **358**, 46–60 (2019). https://doi.org/10.1016/j.cam.2019.03.004.

[40] Filipov, S. M., Gospodinov, I. D. and Faragó, I., Shooting–projection method for two–point boundary value problems, *Applied Mathematics Letters* **72**, 10–15 (2017). https://doi.org/10.1016/j.aml.2017.04.002.

[41] Ha, S. N., A nonlinear shooting method for two–point boundary value problems, *Computers & Mathematics with Applications* **42**, 10–11: 1411–1420 (2001). https://doi.org/10.1016/S0898-1221(01)00250-4.

[42] *Wolfram Language & System Documentation Center. Numerical Solution of Boundary Value Problems (BVP) – shooting method.* https://reference.wolfram.com/language/tutorial/NDSolveBVP.html.

[43] Lambers, J., *Finite difference methods for two–point boundary value problems*, in: MAT 461/561, Spring Semester 2009–10, in: Lecture 26 Notes, Department of Mathematics, University of Southern Mississippi (2010). http://www.math.usm.edu/lambers/mat461/spr10/lecture26.pdf.

[44] Cuomo, S. and Marasco A., A numerical approach to nonlinear two-point boundary value problems for ODEs, *Comput. Math. Appl.* **55**, 2476-2489 (2008). https://doi.org/10.1016/j.camwa.2007.10.002.

In: A Closer Look at the Diffusion Equation
Editor: Jordan Hristov
ISBN: 978-1-53618-330-6
© 2020 Nova Science Publishers, Inc.

Chapter 2

DIFFUSION IN HYPERSONIC FLOWS

H. Berk Gür and Sinan Eyi, PhD
Department of Aerospace Engineering,
Middle East Technical University, Ankara, Turkey

Abstract

In hypersonic flows, air goes into chemical reaction due to high temperature. Therefore, in addition to the Navier-Stokes Equations, chemical reaction equations need to be solved to analyze hypersonic flows. A model may be needed to simulate the diffusion phenomena among chemical species. It is possible to implement Fick's Law of Diffusion as well as Stefan-Maxwell Diffusion Equation. Basically, in Fick's Law of Diffusion, the driving force is the species concentration differences. This method is similar to the Fourier Law of Conduction and defines a diffusion coefficient called Diffusivity. The value of this coefficient is constant for all species in this method. In order to improve the accuracy, more complex models, such as Stefan Maxwell Diffusion Equation, may be used. The driving forces, in this method, are not only the concentration differences but also the interactions of species with each other. Results are evaluated by solving the Navier Stokes and finite-rate chemical reaction equations around the Apollo AS-202 Command module. Eleven species are utilized. As diffusion models, Fick's Law of Diffusion, the Diffusivity calculation with binary collision theory and the Stefan-Maxwell Equation are implemented. Result show that more realistic models may be needed for diffusion flux calculations. Due to its elementary formulation, a rough estimation is possible using the Fick's Law. However, it is possible to improve this elementary formulation using the Binary Diffusion Model.

On the other hand, Stefan-Maxwell Equation gives more detailed results since it uses more accurate formulation. In this study, the performances of these models are compared in hypersonic flow conditions.

Keywords: CFD, hypersonic flow field, diffusion, Fick's law of diffusion, Stefan-Maxwell diffusion equation

1. INTRODUCTION

Hypersonic flows are encountered when velocity of the flow is higher than supersonic which involve complex phenomena that include strong shock waves, high temperature gas effects and chemical reactions [1]. Both experimental and computational methods can be utilized, however, there are major drawbacks in experimental methods due to technological and cost related problems. Recently, computational methods are getting more attention because of their analyses capabilities. Therefore, a large effort has been put to improve the accuracy of such codes for hypersonic flows [2] [3]. In order to simulate these flows, aerothermodynamics discipline which can be defined as combination of thermodynamics, aerodynamics and, chemical reactions should be utilized [4].

Hypersonic flow usually occurs while a spacecraft re-enter the atmosphere. It means that the flow is very fast and the velocity can range between 7 and 12 km/s, which may correspond to a free stream Mach number between 20 to 50 [5]. Also, the re-entry flow field can be divided into 5 different zones. The first zone is called free-molecular zone and the altitude is greater than 120 km. The Knudsen number, which is the ratio of mean free path to characteristic length, is greater than 10, hence the collisions between molecules can be ignored in this region. There are almost no collisions between molecules, and, shock wave may not be formed in this zone. [5].

The second region is called the transition zone which extends from 90 to 120 km in altitude. In this region, the collisions between molecules are more frequent and the Knudsen number ranges between 1 and 0.001. If the altitude is lower than 90 km then the Knudsen number is less than 0.01. This zone is called continuum region, and the interaction between molecule and incoming flow is so high that shock wave may be formed. In this zone, flow characteristics satisfy the continuum approach. In an altitude lower than 80 km, a lower flow speed is observed generally due to the change of kinetic energy to thermal energy. The maximum convective heat rate may be observed in an altitude between 70 and

60 km [5].

Due to the interaction between molecules, in continuum region, strong shock waves occur in front of the capsule [4]. This shock can be characterized by the large velocity, temperature and pressure changes [6]. The region between shock wave and the solid wall is called shock layer [7]. This thin shock layer is one of the characteristic of the hypersonic flows [1]. There are also interactions between viscous boundary layer and shock layer.

The following basic chemical reactions and new species occur in the atmosphere due to high temperature in hypersonic flows. If the temperature ranges between 2000 K to 4000 K the molecular oxygen becomes atomic oxygen [5].

$$O_2 \longrightarrow 2O \quad 2000 < T < 4000 \tag{1}$$

If the temperature is in between 4000 K and 9000 K then Nitrogen starts to dissociate.

$$N_2 \longrightarrow 2N \quad 4000 < T < 9000 \tag{2}$$

Finally, if the temperature is above 9000 K, both Oxygen and Nitrogen start to ionize.

$$\begin{aligned} N &\longrightarrow N^+ + e^- \quad T > 9000 \\ O &\longrightarrow O^+ + e^- \quad T > 9000 \end{aligned} \tag{3}$$

Gosse and Candler claims that, accurate diffusion models are needed to solve Hypersonic Flow regimes since species play an important role when the temperature and velocity increase [8]. Yoon and Rasmussen indicate that multi-component mixtures can react with one another in most hypersonic flow regimes [9]. These reactions play an important role in spacecraft's safety. Desmeuzes, Duffa, Dubroca emphasizes the necessity of diffusion modeling in order to solve forces acting on the body precisely [10]. Diffusion is a physical process which acts on the chemical species. Sutton and Gnoffo underline the importance of diffusion models in heat flux calculations [11].

The simulation of chemically reacting flows requires the solution of a numerically stiff system of nonlinear equations. The stiffness can be caused by the time scale differences in chemical reaction and flow equations. The time scales of chemical reactions may be orders of magnitude faster than that of fluid flow [12]. Step sizes in explicit methods are severely restricted due to this numerical stability problems. Hence, implicit methods are usually preferred for solving these equations. Newtons method is one of the important implicit algorithms for solving systems of nonlinear equations. The main advantage of this method is having quadratic convergence. However, this is only possible if a good initial solution is supplied. Newtons method requires the evaluation of Jacobian matrix that may be somewhat inconvenient and often tedious especially in the solutions of chemically reacting flows. In addition, solving and storing a large Jacobian matrix at each iteration may increase the CPU time and memory requirements, especially in large size problems. Inexact Newton methods developed to keep the advantages and eliminate the disadvantages of Newtons method [13]. At each iteration of these methods, the Newton equation is solved approximately by using an efficient iterative solver like the Jacobian Free Newton Krylov (JFNK) method. As opposed to Newtons method, the JFNK method does not require to explicitly form, store or solve any Jacobian matrices. However, having these advantages, the JFNK method has not been used to a large extent within the CFD community [14]. This may be due to the fact that the efficiency of the JFNK method depends on several factors such as choosing the forcing term [15], implementation of preconditioners [16] and etc. Thus, there are relatively few studies on solving chemically reacting flow and diffusion equations using the JFNK method. Most of these studies are related to the analysis of low speed chemically reacting flows [17], [18]. To the best of the authors' knowledge, no literature is available concerning the implementation of the JFNK method for the analyses of hypersonic flows. In the present study, the three-dimensional Navier-Stokes, finite rate chemical reaction and diffusion equations are solved using the JFNK method. Computations are performed around the Apollo AS-202 command module for the Earth reentry conditions.

2. Physical Modeling

Reynolds Averaged Navier-Stokes (RANS) Equations may not be sufficient to solve hypersonic flow regimes. The addition of Conversation of Species' equations may be needed to solve these flow fields. In this section, the governing equations are explained. However, before going into detail, it is essential to clarify the assumptions used in this study.

The flow domain is assumed to satisfy the continuum approach, so that the Navier-Stokes equations can be implemented. In order to satisfy the continuum approach, the re-entry altitude is taken approximately 40 km above the sea level. Viscous and heat transfer terms are included in the the Navier Stokes solver. The flow domain is solved under the steady-state assumption, so that there are no terms related to time. The other assumptions are given in respective sections.

2.1. Mixtures Properties

In this study, Earth's atmosphere is selected as the free-stream flow domain. However, in hypersonic flow condition, gases around the vehicle consist of different chemical species, ions, electrons and, hence, Earth's atmosphere is treated as a chemically reacting gas mixture. The total density of the mixture is equal to the sum of each species' density as given in Equation 4, which satisfies the continuity equation.

$$\rho = \sum_{i=1}^{Ns} \rho_i \tag{4}$$

Another important parameter is called mass fraction of species.

$$Y_i = \frac{\rho_i}{\rho} \tag{5}$$

The mass fraction is the ratio of species' density to total density. This relation must also satisfy the continuity equation. Therefore, the sum of all mass fractions is equal to unity. The final parameter is called mole fraction which can be defined using the density and molar concentration of species.

$$\rho_i = M_i x_i \tag{6}$$

Mole fraction can be calculated using Equation 6, and it is the ratio of species' molar concentration to total molar concentration.

$$x_i = \frac{M_i}{M} \tag{7}$$

It is easy to show that, the sum of entire species' mole fraction is equal to unity.

2.2. Conservation Equations

There are four conservation equations. In addition to the conservation of momentum equations in x, y and z directions, the conservation of mass, energy and species equations are solved in generalized coordinates.

$$\frac{\partial(\hat{F} - \hat{F}_v)}{\partial \xi} + \frac{\partial(\hat{G} - \hat{G}_v)}{\partial \eta} + \frac{\partial(\hat{H} - \hat{H}_v)}{\partial \zeta} - \hat{S}_{cv} = 0 \tag{8}$$

where the vectors in Equation 8 are defined as follows:

$$\hat{W} = \frac{1}{\mathcal{J}} \begin{bmatrix} \rho \\ \rho u \\ \rho v \\ \rho w \\ \rho E \\ \rho_1 \\ \vdots \\ \rho_{N_s-1} \end{bmatrix}, \hat{S}_{cv} = \frac{1}{\mathcal{J}} \begin{bmatrix} 0 \\ 0 \\ 0 \\ 0 \\ 0 \\ w_1 \\ \vdots \\ w_{N_s-1} \end{bmatrix}, \tag{9}$$

$$\hat{F} = \frac{1}{\mathcal{J}} \begin{bmatrix} \rho \hat{U} \\ \rho u \hat{U} + P\xi_x \\ \rho v \hat{U} + P\xi_y \\ \rho w \hat{U} + P\xi_z \\ (\rho E + P)\hat{U} \\ \rho_1 \hat{U} \\ \vdots \\ \rho_{N_s-1} \hat{U} \end{bmatrix} \tag{10}$$

$$\hat{F}_v = \frac{1}{\mathcal{J}} \begin{bmatrix} 0 \\ \tau_{xx} \\ \tau_{xy} \\ \tau_{xz} \\ \tau_{xx}\hat{U} + \tau_{xy}\hat{V} + \tau_{xz}\hat{W} - q_x + \Sigma J_{x,s} h_s \\ -J_{x,1} \\ \vdots \\ -J_{x,N_s-1} \end{bmatrix} \tag{11}$$

where, u, v and w are the velocity components in the x, y and z directions respectively, ρ is the mixture density and ρ_i is the density of the species i. Also, P is the pressure, E is the total energy. Finally, ω_i is the source term of the species i, and J_{ij} is the diffusion flux between species i and j and $\mathcal{J}$ is Jacobian of coordinate transform.

It can be noted that there are $(N_s - 1)$ Species Continuity Equations. This can be explained with the appearance of the Conservation of Mass equation. The Species Continuity Equations must satisfy the Conservation of Mass Equation, thus the last species' equation is calculated using the Conservation of Mass Equation.

2.3. Thermodynamic and Chemical Models

When the temperature is higher than 800 K, the air molecules are excited in terms of the vibrational energy. As a result, the specific heat of the air becomes temperature dependent, so that the following fourth order polynomials are used to define specific heat.

$$\begin{aligned}\frac{C_{pk}}{R} = a_{1,k}\frac{1}{T^4} + a_{2,k}\frac{1}{T^3} + a_{3,k}\frac{1}{T^2} + a_{4,k}\frac{1}{T} + \\ a_{5,k} + a_{6,k}T + a_{7,k}T^2 + a_{8,k}T^3 + a_{9,k}T^4\end{aligned} \quad (12)$$

The coefficients in Equation 12, are taken from JANAF Tables [19], and the enthalpy and entropy changes can be evaluated as:

$$\begin{aligned} dh_k &= C_{pk}dT \\ ds_k &= C_{pk}\frac{dt}{T} \end{aligned} \quad (13)$$

If the Equation 13 is integrated with respect to temperature, the following equations are obtained;

$$\begin{aligned}\frac{h_k}{R} = -a_{1,k}\frac{1}{3T^3} - a_{2,k}\frac{1}{2T^2} - a_{3,k}\frac{1}{T} + a_{4,k}ln(T) + \\ a_{5,k}T + a_{7,k}\frac{T^3}{3} + a_{8,k}\frac{T^4}{4} + a_{9,k}\frac{T^5}{5} + a_{10,k}\end{aligned} \quad (14)$$

$$\begin{aligned}\frac{s_k}{R} = -a_{1,k}\frac{1}{4T^4} - a_{2,k}\frac{1}{3T^3} - a_{3,k}\frac{1}{2T^2} + a_{4,k}\frac{1}{T} + \\ a_{5,k}ln(T) + a_{6,k}T + a_{7,k}\frac{T^2}{2} + a_{8,k}\frac{T^3}{3} + a_{9,k}\frac{T^4}{4} + a_{11,k}\end{aligned} \quad (15)$$

2.4. Chemical Model

O_2 and N_2 are the main species that can be found in the free stream air. In this study, reacting air assumed to have 11 species, N, O, NO, N_2, O_2, their ionized components and finally electrons.

$$\sum_{k=1}^{K} \nu'_{k,i} X_k \underset{k_{bi}}{\overset{k_{fi}}{\Longleftrightarrow}} \sum_{k=1}^{K} \nu''_{k,i} X_k \quad (16)$$

For I number of reactions and K number of species stoichiometric reactions can be written as in Equation 16. The terms $\nu'_{k,i}$ and $\nu''_{k,i}$ are called stoichiometric coefficients of reactants and products in reaction I. Also, X_k is a notation for k^{th} species. Finally, the notations k_{fi} and k_{bi} are used for forward and backward reaction rates.

$$k_{fi} = A_i T^{\beta_i} \exp\left(\frac{-E_i}{R_u T}\right) \quad (17)$$

For forward reaction rate calculations, Arrhenius-type of equation is utilized as in Equation 17, where A_i is the pre-exponential factor and E_i is the activation energy.

2.5. Diffusion Models

The diffusion fluxes are utilized in Species' Continuity Equations and Conservation of Energy Equation. It is convenient to define the non-dimensional numbers related to diffusion flux calculations.

$$\begin{array}{c} Sc = \frac{\nu}{D} = \frac{\mu}{\rho D} \\ Le = \frac{\alpha}{D} \\ Pe = \frac{Lu}{D} = ReSc \end{array} \quad (18)$$

Schmidt Number (Sc) is the ratio of viscous diffusion (μ) to molecular diffusion (D), Lewis Number (Le) is the ratio of thermal diffusivity (α) to molecular diffusivity, Peclet Number (Pe) is the ratio of advective transport rate to diffusive transport rate, and finally Re stands for Reynolds number in the equation above.

One of the simplest model is Fick's Law of diffusion. In this law, the diffusivity of the species relies only on the concentration differences.

$$J_i = -\rho D_{ij} \nabla Y_i \quad (19)$$

Fick's Law of Diffusion can be written in terms of the mass fraction of species as presented in Equation 19. Furthermore, it should be noted that the equation is written for two different species. The Fick's Law of Diffusion can be modified to solve more than two species as follows:

$$J_i = -\rho \sum_{j=1}^{Ns-1} D_{ij} \nabla Y_i \quad (20)$$

In equation above, the term D_{ij} is called diffusion constant or diffusivity in short. This term depends on species "i" and "j", therefore, in order to define it, two or more different species are needed. In Fick's Law of Diffusion, this term is becomes invalid when "i" is equal to "j", which means that, there is no self diffusion.

The diffusivity is calculated for every single species where collision occurs simultaneously between them. In general, the value of diffusivity can be determined from binary collision theory which evaluates the diffusion constant for species pairs in the mixture. The diffusion constant can be evaluated as follows:

$$D_{ij} = \frac{\overline{D}}{F_i F_j} \quad (21)$$

In Equation 21, F_i and F_j are calculated from the molecular weight of the species where, $\overline{D}$ is the self diffusion coefficient which, can be written as a function of pressure and temperature, since diffusion is effected by these variables.

$$F_i = (\frac{M_i}{26})^{0.461} \quad (22)$$

$$\overline{D} = \frac{CT^{1.5}}{P} \quad (23)$$

where, C is a constant and P is the pressure. The self-diffusion coefficient can be calculated aside from [21]. It can be evaluated using with a correlation [23].

$$\overline{D} = 2.69280 x 10^{-3} \frac{T(\frac{T}{M_{ref}})^{1/2}}{p\Omega_{ref}^2 \sigma_{ij}^{(1,1)*}} \quad (24)$$

In the Equation above, $\sigma_{ij}^{(1,1)*}$ is an integral in transport property which, depends on the particular intermolecular potential function [22]. This property can be calculated by the Lennard-Jones Potential Function. This is a mathematical approach that can simply approximate the interaction between species. In this method, Lennard-Jones Potential helps to model the effect of species.

$$D_{ij} = 1.858 x 10^{-3} \frac{(T^3(\frac{1}{M(i)} + \frac{1}{M(j)}))^{0.5}}{P\sigma_{ij}^2 \Omega_{ij}} \quad (25)$$

where, Ω_{ij} is calculated with the Lennard-Jones Potential, and σ_{ij} is related to the species. Also, $M(i)$ and $M(j)$ is the molecular weight of 'i'th and 'j'th species.

$$\Omega_{ij} = 0.65 - 1.33 * 10^{-3}\frac{T}{\epsilon_{ij}} + 0.7903(\frac{T}{\epsilon_{ij}})^{-0.8} \qquad (26)$$

$$\sigma_{ij} = \frac{\sigma_i + \sigma_j}{2}$$
$$\epsilon_{ij} = \sqrt{\epsilon_i \epsilon_j} \qquad (27)$$

The values of the σ and ϵ for different species are given in Table 1.

Table 1. Epsilon and Sigma values

Species	σ(Angstroms)	$\epsilon\ k_b$ (K)
O	2.75	80
N	3.298	71.04
N_2	3.621	97.53
O_2	3.458	107.400
NO	97.530	3.621

Stefan-Maxwell Diffusion Equation can be used as an alternative model to evaluate the diffusion flux. In Fick's model, the driving forces of the diffusion is the species' concentration differences. For more realistic calculations, Stefan-Maxwell Diffusion Equation can be used. Originally, the Stefan-Maxwell Diffusion Equation is developed to solve the mole fraction gradient.

$$\nabla x_i = \frac{M}{\rho}\sum_{j\neq i}^{Ns}(\frac{x_i J_j}{M_j D_{ij}} - \frac{x_j J_i}{M_i D_{ji}}) \qquad (28)$$

Equation 28 can be re-arranged to solve diffusion fluxes instead of mole fraction gradient.

$$J_i = -\rho\frac{M_i}{M}\frac{D_{im}}{(1-x_i)}\nabla x_i + \frac{Y_i}{(1-x_i)}D_{im}\sum_{j\neq i}^{Ns}\frac{MJ_j}{M_j D_{bij}} \qquad (29)$$

In this study, mass fraction is taken as the main variable in the diffusion flux calculations, so that, Equation 29 can be written in terms of mass fractions.

$$J_i = -\rho D_{im} \nabla Y_i + \frac{Y_i}{(1-x_i)} D_{im} \sum_{j\neq i}^{Ns} (\rho \frac{M}{M_j} \nabla Y_j + \frac{M}{M_j}\frac{J_j}{D_{bij}}) \quad (30)$$

As seen from Equation 30, the first part of the right hand side represents Fick's Law of Diffusion, which takes the concentration differences of the species. In the second part of the right hand side, the other species diffusion fluxes are included to calculate diffusion flux of species "i".

In Stefan-Maxwell Equation, the calculation of diffusivity is different from Ficks Law because diffusivity is assumed as a constant in Ficks Law.

$$D_{ij} = D_{bij}[1 + \frac{x_k(\frac{M_k}{M_j}D_{bik} - D_{bij})}{x_i D_{bjk} + x_j D_{bik} + x_k D_{bij}}] \quad (31)$$

The calculation of diffusivity depends on mole fractions, and D_{bij} which is called binary diffusion coefficient. This equation can be modified again for 11 species calculation.

$$\rho D_{bij} = 7.1613x10^{-25}\frac{M[T(\frac{1}{M_i} + \frac{1}{M_j})]^{0.5}}{\Omega_{ij}} \quad (32)$$

The collision is one of the important phenomena in diffusion, so that collision cross section for mass diffusion, Ω, has to be calculated.

$$\Omega_{ij} = \pi(r_i + r_j)^2 \quad (33)$$

For example, the diffusivity for oxygen and nitrogen is different from oxygen and hydrogen. The question is how one can calculate the diffusion flux of oxygen and which diffusivity term can be used?

$$D_{i,eff} = \sum_{j=1}^{N_s-1} D_{ij}\frac{\Delta x_j}{\Delta x_i} \quad (34)$$

This question can be answered by defining the effective diffusivity as in Equation 34.

3. MATHEMATICAL MODELING

In the solution of flow equations, a cell centered finite volume method has been used. Then, the governing equation can be discretized as;

$$\begin{aligned}
& (\hat{F}_c - \hat{F}_v)_{i+\frac{1}{2},j,k} - (\hat{F}_c - \hat{F}_v)_{i-\frac{1}{2},j,k} + \\
& (\hat{G}_c - \hat{G}_v)_{i,j+\frac{1}{2},k} - (\hat{G}_c - \hat{G}_v)_{i,j-\frac{1}{2},k} + \\
& (\hat{H}_c - \hat{H}_v)_{i,j,k+\frac{1}{2}} - (\hat{H}_c - \hat{H}_v)_{i,j,k-\frac{1}{2}} - \\
& \hat{S}_{i,j,k} = 0
\end{aligned} \quad (35)$$

In equation 35, $i \pm \frac{1}{2}$, $j \pm \frac{1}{2}$ and $k \pm \frac{1}{2}$ represent the cell interfaces. Also, for the spatial discretization of the flux vectors, upwind flux splitting schemes are used.

$$\begin{aligned}
& [\hat{F}_c^+(\hat{Q}^-)_{i+\frac{1}{2},j,k} + \hat{F}_c^-(\hat{Q}^+)_{i-\frac{1}{2},j,k}] - [\hat{F}_c^+(\hat{Q}^-)_{i-\frac{1}{2},j,k} + \\
& \hat{F}_c^-(\hat{Q}^+)_{i-\frac{1}{2},j,k}] + [\hat{G}_c^+(\hat{Q}^-)_{i,j+\frac{1}{2},k} + \hat{G}_c^-(\hat{Q}^+)_{i,j-\frac{1}{2},k}] - \\
& \hat{G}_c^+(\hat{Q}^-)_{i,j-\frac{1}{2},k} + \hat{G}_c^-(\hat{Q}^+)_{i,j-\frac{1}{2},k} + \hat{H}_c^+(\hat{Q}^-)_{i,j,k+\frac{1}{2}} + \\
& \hat{H}_c^-(\hat{Q}^+)_{i,j,k-\frac{1}{2}} - \hat{H}_c^+(\hat{Q}^-)_{i,j,k-\frac{1}{2}} + \hat{H}_c^-(\hat{Q}^+)_{i,j,k-\frac{1}{2}} - \\
& \hat{S}_{i,j,k} = 0
\end{aligned} \quad (36)$$

There are several different flux splitting methods like Steger-Warming, van Leer, and AUSM schemes. In this study, only van Leer Flux Splitting method is used. In this method, Mach number is used to split the flux vector.

$$M = M^+ + M^- \quad (37)$$

The following relations are utilized in the evaluation of M^+ and M^-.

$$M^+ = \begin{cases} \frac{1}{4}(M+1)^2 & |M| \leqslant 1 \\ \frac{1}{2}(M+|M|) & otherwise \end{cases}$$
$$M^- = \begin{cases} -\frac{1}{4}(M-1)^2 & |M| \leqslant 1 \\ \frac{1}{2}(M-|M|) & otherwise \end{cases} \quad (38)$$

Diffusion in Hypersonic Flows

The Mach Number splitting can be used for both subsonic and supersonic flows. If the flow is supersonic then the overall scalar values of Mach number are directed to downstream.

$$\hat{F}_c^{\pm} = \pm \frac{\rho a}{\mathcal{J}} \frac{(M+1)^2}{2} \begin{bmatrix} 1 \\ \frac{1}{\gamma}(-\hat{U}_\xi \pm 2a)\hat{\xi}_x + u \\ \frac{1}{\gamma}(-\hat{U}_\xi \pm 2a)\hat{\xi}_y + u \\ \frac{1}{\gamma}(-\hat{U}_\xi \pm 2a)\hat{\xi}_z + u \\ (\frac{-\hat{U}_\xi \pm 2a}{\gamma + 1})\hat{U}_\xi + \frac{2a^2}{\gamma^2 - 1} + \frac{u^2 + v^2 + w^2}{2} \\ \frac{\rho_1}{\rho} \\ \vdots \\ \frac{\rho_{Ns-1}}{\rho} \end{bmatrix}$$
(39)

where, $\mathcal{J}$ is the Jacobian Matrix, a is the speed of sound and γ is the specific heat ratio.

In CFD applications, it is important to define the correct boundary conditions. There are mainly three boundary conditions; far field, wall and symmetry boundaries. The suitable boundary conditions can be obtained by using ghost cells. The far-field boundary conditions should be defined according to the direction of the information coming from.

The wall boundary condition is defined as no mass or energy goes into the wall. As mentioned in the Physical Modeling section, Navier-Stokes Equations are used to solve the flow domain. One of the properties of Navier-Stokes Equation is that no velocity component on the wall. Therefore, all velocity components on the wall are zero, whereas the other flow variables are extrapolated from interior cells. For energy, the adiabtaic wall boundary condition is implemented at the wall. In order to save some computational cost, half of the domain is calculated due to the symmetry of the capsule. The flow variables are equal to each other on the symmetry plane except for the velocities normal to the plane. These velocities have the opposite directions.

In Newton's method the residual equation can be written as follows:

$$\hat{R}(\hat{W}) = \frac{\partial \hat{F}(\hat{W})}{\partial \xi} + \frac{\partial \hat{G}(\hat{W})}{\partial \eta} + \frac{\partial \hat{H}(\hat{W})}{\partial \zeta} - (\frac{\partial \hat{F}_v(\hat{W})}{\partial \xi} + \frac{\partial \hat{G}_v(\hat{W})}{\partial \eta} + \frac{\partial \hat{H}_v(\hat{W})}{\partial \zeta}) - \hat{S} = 0 \quad (40)$$

The magnitude of residual vector, $\hat{R}(\hat{W})$, in Equation 40 determines the accuracy of the solution. The Taylor expansion can be applied for Equation 40.

$$\hat{R}^{n+1}(\hat{W}) = \hat{R}^n(\hat{W}) + (\frac{\partial \hat{R}}{\partial \hat{W}})^n \Delta \hat{W}^n \quad (41)$$

If the flow variables are assumed to satisfy residual equations exactly at the iteration of n+1, Equation 41 can be organized as follows:

$$(\frac{\partial \hat{R}}{\partial \hat{W}})^n \Delta \hat{W}^n = -\hat{R}^n(\hat{W}) \quad (42)$$

where, $\frac{\partial \hat{R}}{\partial \hat{w}}$ is the Jacobian matrix and it is solved for the increment in state variable vector, $\hat{W}$. In Newton's method, converged solutions can be achieved with a small number of iterations. However, forming and solving large Jacobian matrices at each iteration is one the disadvantages of this method. Requiring a good initial solution is another disadvantage of this method. In order to eliminate the disadvantages, of Newton's method, a new approach has been implemented and JFNK method is developed. In this approach, the following equation is utilized to solve Newton's method approximately at each iteration.

$$\|\hat{R}^n(\hat{W}) + (\frac{\partial \hat{R}}{\partial \hat{W}})^n \Delta \hat{W}^n\| = \mu \|\hat{R}^n(\hat{W})\| \quad (43)$$

In Equation 43, the term μ is the forcing term and varies between zero and one. The value of this term determines the convergence characteristics of the JFNK.

4. Results And Discussion

In this study, Earth's atmosphere is selected as the free-stream domain. As mentioned in the physical modeling section, the flow must satisfy the continuum approach, therefore, the altitude is selected as 40 km above sea level.

Diffusion in Hypersonic Flows

Table 2. Free stream conditions

Parameter	Value
Temperature (K)	220
Pressure (kPa)	664
Mach Number	10.18
Reynolds Number	1.1×10^6
Angle of Attack (^{0})	0
Species	$11(O_2, N_2, N, O, NO, O_2^+, N_2^+, N^+, O^+, NO^+, e^+)$

The free stream conditions are summarized in Table 2. Navier Stokes and finite-rate chemical reaction equations are solved around the Apollo AS-202 Command module. Since the capsule is symmetric, half of the flow domain is analyzed. In this way, the computational time can be reduced significantly.

Table 3. List of cases used in this study

Case Number	Diffusivity	Equation
1	CLN	Fick's Law
2	BCT	Fick's Law
3	SMDE	Stefan-Maxwell

The list of cases is summarized in Table 3. In diffusivity calculation three models are implemented. CLN is Constant Lewis Number model with $Le = 1.4$, BCT is Binary Collision Theory model, and $SMDE$ is Stefan-Maxwell diffusivty equation.

4.1. Grid Independence Study and Validation

It is essential to define the correct grid size before any analyses. A grid independent study is performed to find a suitable grid size. The main goal in this study is to determine a grid size which has a low computational time yet, the results must be almost independent from the finer grid size. Four different grid sizes are selected to test the grid independence.

In Table 4, grid sizes and the computational times are given. The grid independence study is performed for the Fick's Law of Diffusion case. It can be

Table 4. Four different grid sizes and computational time

Grid Name	ixjxk	Number of Cells	Time (s)
Grid-1	64x32x17	34816	238300
Grid-2	96x48x26	86112	736012
Grid-3	112x56x30	188160	1183129
Grid-4	128x64x34	278528	1896158

noted that, the computational time is increasing with the increasing grid size. The convergence characteristics with each grid are studied in Figure 1. The iterations are stopped when the norm residual reaches 10^{-5}. The non-dimensional pressure distribution evaluated from different grids are shown in Figure 2. Although computational results are in good agreement with experimental data in all cases, the finest grid is used for diffusion analyses.

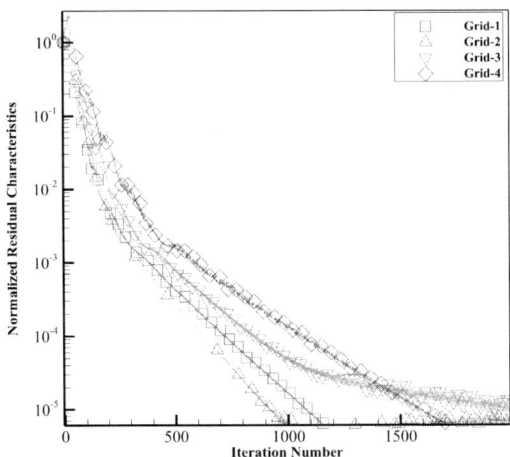

Figure 1. Residual characteristics of four different grids.

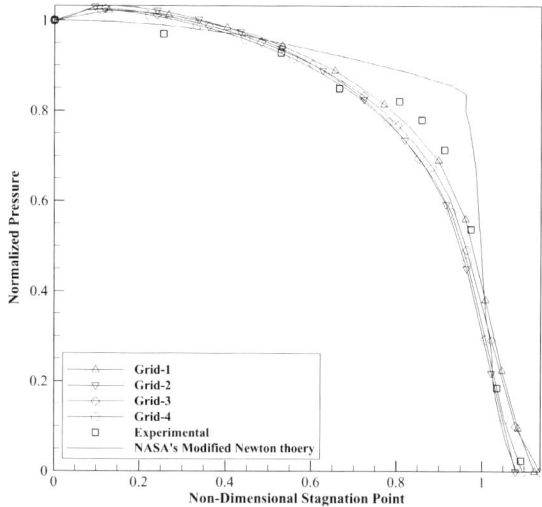

Figure 2. Non-dimensional pressure distribution in stagnation point.

4.2. Results of Fick's Law of Diffusion

First, Fick's Law of Diffusion with Constant Lewis Number Model is implemented to the code, and the solutions with and without diffusion models are compared.

Figure 3 shows the temperature distribution along the stagnation line. Shock locations and temperature changes predicted by both models are very close to each other. However, behind the shock wave, higher temperature is observed with the diffusion model. In this region, the difference in temperature evaluated with and without diffusion model is around 300 K. This is an important temperature difference, hence it directly affects the design of the heat shield.

One of the major species in the atmosphere is oxygen. The mass fraction distributions of molecular oxygen without any diffusion model and with Fick's Law of diffusion are shown in Figure 4. Since the temperature approaches to 4000 K, the mass fraction of molecular oxygen decreases because of dissociation and new species are produced. However, behind the shock wave, temperature decreases slightly and oxygen atoms recombine to produce molecular oxygen again. The differences in mass fraction with and without diffusion model can be clearly seen near the stagnation point. Another major species in the air

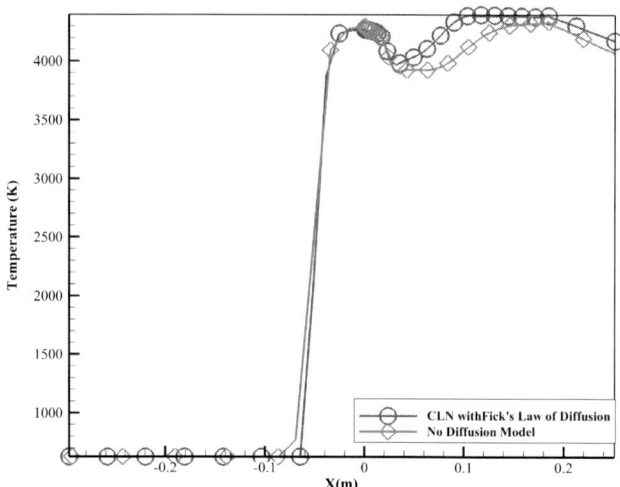

Figure 3. Temperature distribution along the stagnation line with CLN with Fick's Law.

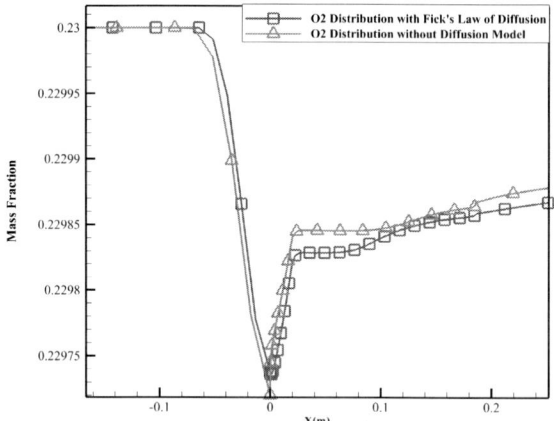

Figure 4. O2 distribution along stagnation line with and without diffusion model.

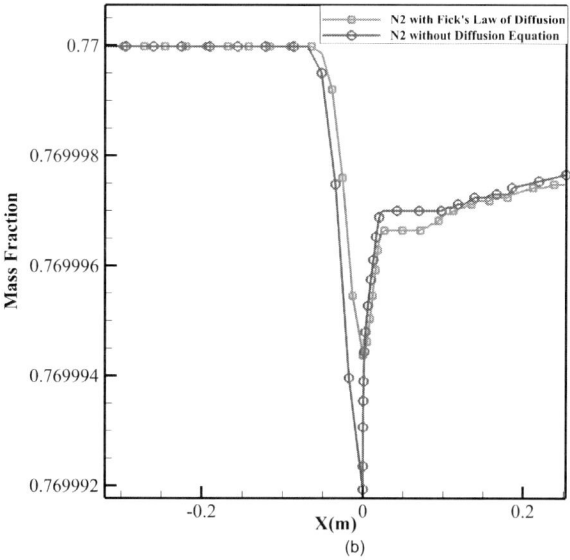

Figure 5. N2 distribution along stagnation line with and without diffusion model.

is nitrogen. The mass fraction variation of molecular nitrogen with and without diffusion model is shown in Figure 5.

4.3. Results of Binary Collision Theory

In order to improve Fick's Law of Diffusion, Binary Collision Theory (BCT) is implemented to calculate diffusivity. As it was mentioned in the physical modeling section, Diffusivity is assumed to be constant in Ficks Law. However, it must be calculated in BCT. In Figure 6, the results from the present study and literature [23] are compared and generally good agreements are achieved.

The variations of temperature along the stagnation line without any diffusion model and with CLN and BCT models are compared in Figure 7. The differences become more significant behind the shock wave. Figures 8 and 9 show the similar differences in the mass fractions of O_2 and $N2$ evaluated with and without diffusion models.

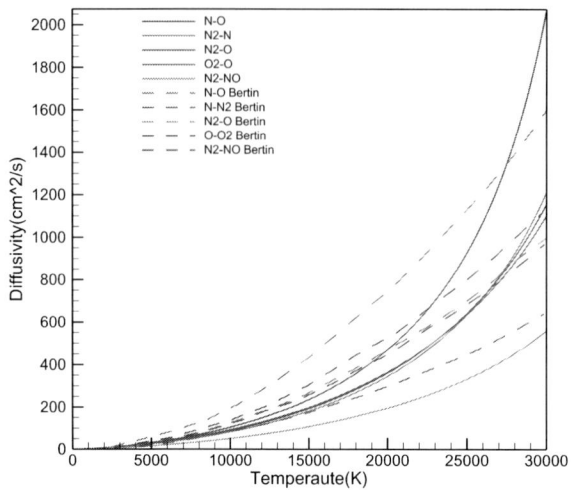

Figure 6. Comparison between calculated diffusivity with literature [23]

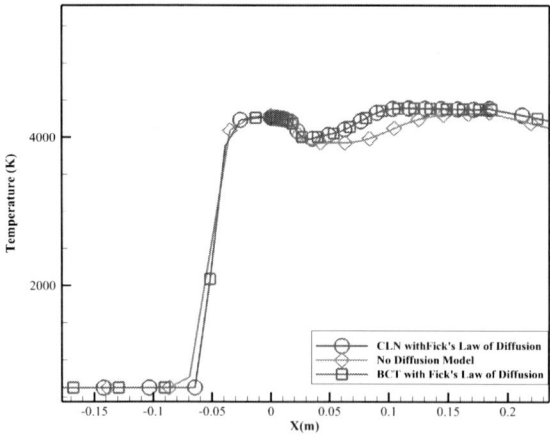

Figure 7. Temperature distribution along the stagnation line with CLN with Fick's Law and BCT with Fick's Law.

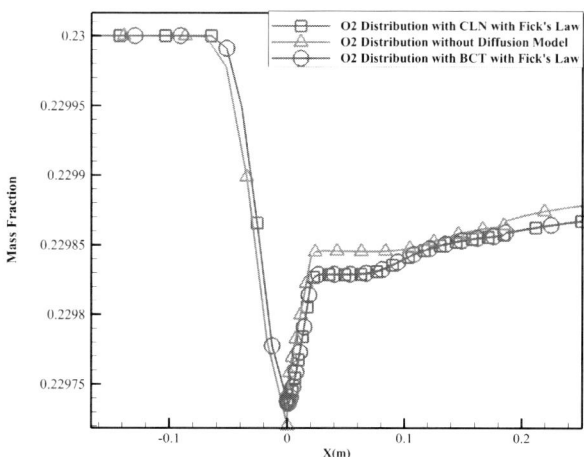

Figure 8. O2 distribution along the stagnation line with CLN with Fick's Law and BCT with Fick's Law.

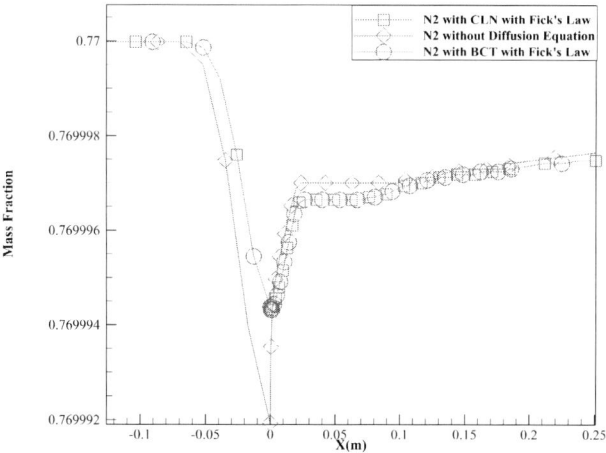

Figure 9. N2 distribution along the stagnation line with CLN with CLN with Fick's Law and BCT with Fick's Law.

4.4. Results of Stefan Maxwell Equation

Stefan-Maxwell Diffusion Equation is implemented to the code as a second diffusion model. By using different diffusion models, the normalized pressure distributions along the normalized wall distance are compared in Figure 10. Near the stagnation point there are a small differences. As can be seen from the figure, away from the stagnation region, Ficks Law predicts pressure closer to the experimental data [22]. However, in the stagnation region where all chemical reactions occur, the Stefan-Maxwell prediction and experimental data in a better agreement.

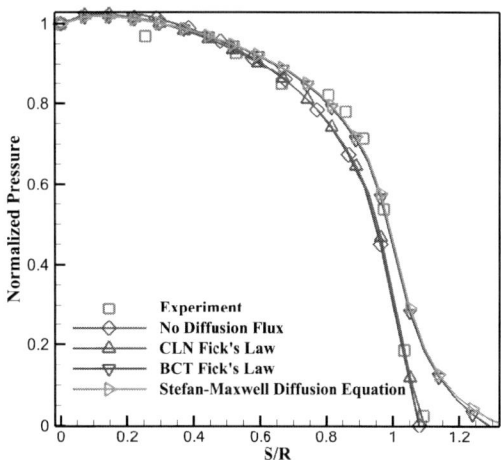

Figure 10. Non-dimensional pressure distribution around the stagnation point.

Since there is almost no differences between CLN and BCT results, from now on, only CLN results will be presented. In Figure 11, the stagnation line temperature distributions predicted using Fick's Law and Stefan Maxwell equation are compared. Behind the shock wave, the temperature calculated using the Stefan Maxwell Equation is lower than the one evaluated with Fick's Law. Using diffusion models produces larger temperature behind the shock wave. The mass fraction distribution of O2 and N2 are evaluated with and without diffusion models and results are shown in Figure 12 and 13. In front of shock wave, the mass fractions of O2 and N2 are lower if no diffusion model is used. However,

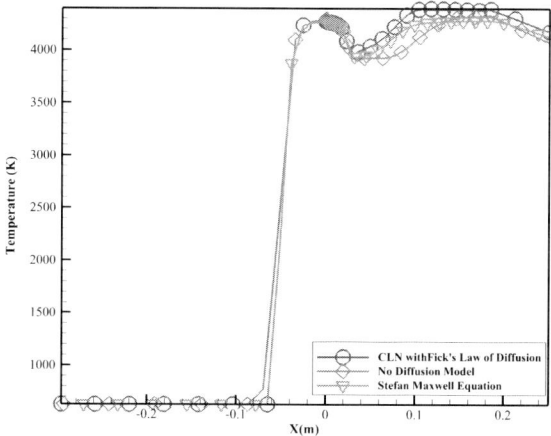

Figure 11. Temperature distribution on the stagnation line with diffusion models.

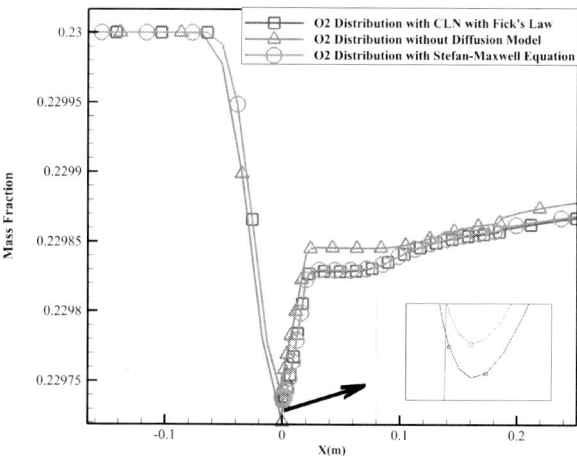

Figure 12. O2 distribution on the stagnation line with diffusion models.

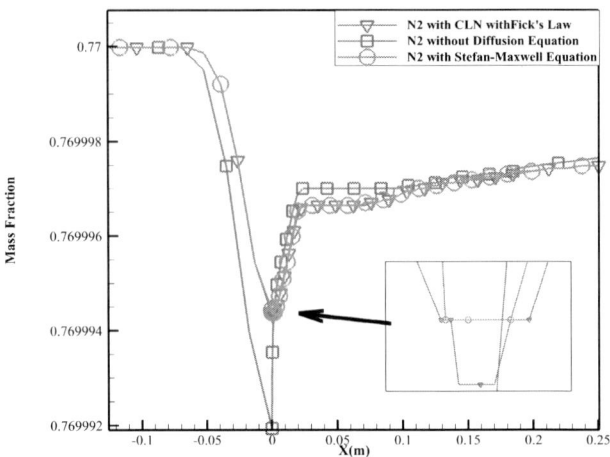

Figure 13. N2 distribution on the stagnation line with diffusion models.

behind the shock wave, the situation becomes opposite.

CONCLUSION AND FUTURE WORK

An in-house CFD code is developed to analyze hypersonic flows using different diffusion models. Fick's Law and the Stefan-Maxwell Diffusion Equation are implemented using Binary Collision Theory for Diffusivity calculation. Although using diffusion models increases the fidelity of the hypersonic flow analyses, it may also increase the computational cost because of extra calculations are needed for diffusion evaluations. Solving Stefan Maxwell equation is computationally more intensive compared to Fick's Law of diffusion.

In order to improve the accuracy of Fick's Law, diffusivity calculations can be done by using Binary Collision Theory. The effects of binary collision theory on diffusion calculation are more significant at high temperature.

The pressure distributions near the stagnation point using Stefan-Maxwell Diffusion equation agree with the experimental data. However, away from the stagnation point, the model over-predicts the pressure distribution. Also, this model does not predict the maximum temperature as high as Fick's Law of Diffusion. Therefore, the selection of diffusion equation depends on the location of the analysis.

Finally, as a future work, more experimental data will be searched to compare the accuracy of diffusion models. Effects of grid sizes on diffusion evaluations will be studied in detail. The convergence characteristics of different diffusion models will be analyzed. Also, the convergence of the calculations will be improved. In addition, the code will run in parallel so that the computational time will be reduced significantly.

REFERENCES

[1] Anderson Jr, J. D. *Fundamentals of Aerodynamics*, Tata McGraw-Hill Education 2010.

[2] Lofthouse A. and Boyd I.D., *Hypersonic flow over a flat plate: CFD comparison with experiment*, 47th AIAA Aerospace Sciences Meeting including The New Horizons Forum and Aerospace Exposition, 1315 (2009).

[3] Padilla J. F. and Boyd I.D., *Assessment of Gas-surface Interaction Models for Computation of Rarefied Hypersonic Flow*, Journal of Thermophysics and Heat Transfer **23(1)**, 96-105 (2009).

[4] Subrahmanyam P., *Development of an Interactive Hypersonic Flow Solver Framework for Aerothermodynamic Analysis*, Engineering Applications of Computational Fluid Mechanics **2(4)**, 436-455 (2008).

[5] Scalabrin L. C., *Numerical Simulation of Weakly Ionized Hypersonic Flow over Reentry Capsules*, PhD thesis, University of Michigan 2007.

[6] Jain S. U., *Hypersonic Non-equilibrium Flow Simulation over a Blunt Body using BGK Method*, Texas A and M University 2007.

[7] Mathews R.N. and Shafequee A. P., *Hypersonic Flow Analysis on an Atmospheric Re-entry Module*, International Journal of Engineering Research and General Science **3(5)**,991-1001 (2015).

[8] Gosse, R. and Candler G., *Diffusion Flux Modeling: Application to Direct Entry Problems*, 43rd AIAA Aerospace Sciences Meeting and Exhibit., 389 (2005).

[9] Yoon B. and Rasmussen M. L., *Diffusion Effects in Hypersonic Flows with a Ternary Mixture*, KSME International Journal **13(5)**,432-442 (1999).

[10] Desmeuzes C., Duffa G. and Dubroca B., *Different Levels of Modeling for Diffusion Phenomena in Neutral and Ionized Mixtures*, Journal of thermophysics and heat transfer **11(1)**, 36-44 (1997).

[11] Sutton K. and Gnoffo P., *Multi-component Diffusion with Application to Computational Aerothermodynamics*, 7th AIAA/ASME Joint Thermophysics and Heat Transfer Conference, 2575 (1998).

[12] LeVeque R.J. anb Yee, H. C., *Study of Numerical Methods for Hyperbolic Conservation Laws with Stiff Source Terms*, Journal of Computational Physics **86**, 187-210 (1990).

[13] Dembo R.S., Eisenstat S.C. and Steihaug T., *Inexact Newton Methods*, SIAM Journal of Analysis **19(2)**, 400-408, (1989).

[14] Chisholm T.T. and Zngg D.W., *A Jacobian-free Newton-Krylov Algorithm for Compressible Turbulent Fluid Flows*, Journal of Computational Physics **228(9)**, 3490-3507 (2009).

[15] Eisenstat S.C. and Walker H.F., *Choosing the Forcing Terms in an Inexact Newton Method*, SIAM Journal on Scientific Computing **17(1)**, 16-32 (1996).

[16] Persson P.O. and J. Peraire J., *Newton-GMRES Preconditioning for Discontinuous Galerkin Discretizations of Navier-Stokes Equations*, SIAM Journal on Scientific Computing **30(6)**, 2709-2733 (2008).

[17] Knoll D. A., Mchugh P.R. and Keyes D.E., *Newton-Krylov Methods for Low-Mach-Numaber Compressible Combustion*, AIAA Journal **34(5)**, 961-967 (1996).

[18] Mchugh P., Knoll D. and Keyes D. *Application of Newton-Krylov-Schwarz Algorithm to Low-Mach-Number Compressible Combustion*, AIAA Journal **26(2)**, 290-292 (1998).

[19] McBride B. J., Zehe M. J., and Sanford G., *Glenn coefficients for calculating thermodynamic properties of individual species*, NASA/TP **211556** 2002.

[20] Kendall R. M., *An analysis of the coupled chemically reacting boundary layer and charring ablator. Part 5-A general approach to the thermochemical solution of mixed equilibrium-nonequilibrium, homogeneous or heterogeneous systems*, National Aeronautics and Space Administration 1968.

[21] Bertin J. J., *The Effect of Protuberances, Cavities, and Angle of Attack on the Wind-tunnel Pressure and Heat-Transfer Distribution for the Apollo Command Module*, National Aeronautics and Space Administration 1966.

[22] Bartlett E. P., Kendall R. M. and Rindal R. A., *An analysis of the coupled chemically reacting boundary layer and charring ablator. Part 4-A unified approximation for mixture transport properties for multicomponent boundary-layer applications*, National Aeronautics and Space Administration 1968.

[23] Bertin J. J., *Hypersonic Aerothermodynamics*, AIAA 1994.

In: A Closer Look at the Diffusion Equation
Editor: Jordan Hristov

ISBN: 978-1-53618-330-6
© 2020 Nova Science Publishers, Inc.

Chapter 3

ON THE NONLINEAR DIFFUSION WITH EXPONENTIAL CONCENTRATION-DEPENDENT DIFFUSIVITY: INTEGRAL-BALANCE SOLUTIONS AND ANALYZES

Jordan Hristov[*]
Department of Chemical Engineering,
University of Chemical Technology and Metallurgy, Sofia, Bulgaria

Abstract

The chapter address analysis and approximate solutions of a diffusion equation with concentration dependent diffusivity of exponential type frequently encountered in polymers and soils. This is a well-known problem solved by various approximate methods. The present chapter applies the integral-balance approach in two versions: Heat-balance integral (HBIM) and Double-integration method (DIM). Analyzes of the solution behaviours with Dirichlet boundary condition and under assumptions of a slow diffusion process are the principle tasks of the study.

Keywords: diffusion, exponential diffusivity, concentration-depended diffusivity, integral- balance method

---

[*]Corresponding Author's Email: jordan.hristov@mail.bg.

1. INTRODUCTION

1.1. Physical Origins of the Modelled Problem

Nonlinear gradient-driven diffusion of fluid and gases are encountered in many applications such as removal or addition of solvents in concentrated polymer solutions [1, 2, 3], wood drying and wetting [4, 5], concrete wetting [6], polymer devolatization [7], drug release from polymer matrices [8], moisture diffusion in porous materials such as reinforced polymeric materials [9, 10], infiltration in soils [11, 12], etc.

The common approach in modelling diffusion (sorption) in polymers, for instance, is based on the Ficks law and when, for instance, a transient diffusion in polymer solution above the glass transition temperature takes place this is still valid [13]. However, reducing the temperature below the glass transient point the diffusion process becomes non-Fickian due to strong effect of the solution viscoelasticity [13, 14, 15]. In fact, local accumulations of solvent during diffusion cause local dilations of the polymer components and due to the delayed response of the polymer structure to such disturbances the local behaviour of the solvent diffusion and the macroscopic performance of the process are affected.

The concentration dependence of the diffusivity is the major phenomenon controlling the flux of the penetrant through the polymer (sorption) [1, 7, 16, 17], drug release [8] diffusion in wood-fiber composites [9, 10], water wetting of porous media (soils)[11, 12], etc. Commonly this relationship can be expressed (as a successful fit of experimental data; see the comments further in the text) in the form

$$D(C) = D_0 \exp(\beta C) \qquad (1)$$

The diffusivity of the medium at zero concentration is D_0, (precisely at a given temperature and an initial concentration C_i) while the rate-factor β and its range of variation depend on the physics of the modelled problem (see section 1.4): when solvents diffuse through polymers this is the so-called *plasticizing coefficient* expressing the effect of the plasticizing action of the penetrant (liquid or gas) on the segmental motions of the polymer. In soil infiltration [11, 12] it has no specific name and is mentioned only as a coefficient with a dimension $[concentration^{-1}]$ or as a dimensionless rate-factor (as it is done in this chapter).

1.2. Mathematical Problem Formulation and Governing Equations

This chapter addresses approximate analytical solutions of one-dimensional diffusion problem subjected to the Dirichlet boundary condition, namely

$$\frac{\partial C}{\partial t} = \frac{\partial}{\partial x}\left(D(C)\frac{\partial C}{\partial x}\right) \qquad (2)$$

$$C(0,t) = C_s, \quad t \geq 0, \quad C(x,0) = C_\infty \qquad (3)$$

The diffusivity is an exponential function of the concentration having the form

$$D(C) = D_0 e^{\beta u}, \quad u = \frac{C - C_\infty}{C_s - C_\infty}, \quad D_0 > 0, \quad \beta > 0 \qquad (4)$$

If the medium is considered as semi-infinite, i.e. *there is no a characteristic length scale of the process* we may accept without a loss of generality that $C_\infty = 0$. Then, with the dimensionless variable $u = C/C_s$ (4) and the relevant boundary and initial conditions the model (2)-(3) can be expressed in a semi-dimensionless form (with respect to u only, not with respect to the time and spatial coordinate) as

$$\frac{\partial u}{\partial t} = \frac{\partial u}{\partial x}\left(D_0 e^{\beta u} \frac{\partial u}{\partial x}\right) \qquad (5)$$

$$u(0,t) = 1, \quad u(\infty, t) = u_\infty = C_\infty/C_s, \quad t \geq 0 \qquad (6)$$

$$u(x,0) = u_\infty, \quad 0 \leq x \leq \infty \qquad (7)$$

Otherwise, when the medium thickness is finite, i.e., there is a characteristic length scale, by introduction of $z = x/L$ and $\tau = t/t_0$ where $t_0 = L^2/D_0$, the model can be expressed in a complete dimensionless form as

$$\frac{\partial u}{\partial \tau} = \frac{\partial u}{\partial z}\left(e^{\beta u} \frac{\partial u}{\partial z}\right) \qquad (8)$$

$$u(0,t) = 1, \quad u(1,t) = u_\infty, \quad t \geq 0 \qquad (9)$$

$$u(x,0) = u_\infty, \quad 0 \leq z \leq 1 \qquad (10)$$

Equation (2) has been studied for diffusivities in various functional forms of $D(u)$ and the power-law diffusivity $D(u) = D_0 u^m$ with $m > 0$ for both slow diffusion ($m > 1$) [21, 22, 23] and fast diffusion ($0 < m < 1$) [21]. It is a special case of the exponential relationship for low u: for this specific case eq. (2) is commonly termed as porous media equation (PME) and it is extensively studied [24].

1.3. The Diffusivity Functional Relationship: Preliminary Comments

For a typical diffusion problem with a Dirichlet boundary condition ($C = C_s$ or $u(x,t) = u(z,\tau) = 1$) and $C_\infty(\infty, t) = 0$ and the case of slow diffusion ($m > 0$) the solution of PME with a power-law diffusivity exhibits a sharp front [22, 23] propagating with a speed preserving the Boltzmann scaling, i.e. the proportional to $t^{1/2}$. Beyond the front the solution is zero. When a slow diffusion is modelled ($m > 1$) there is discontinuity in the first derivative u_x at the front, while for the fast diffusion ($m < 1$) the concentration distribution approaches the x axis smoothly [20]. As it is commented by [20], the PME with a power-law diffusivity exhibits a degenerate behaviour while in the case of exponential diffusivity the condition $D(0) = 0$ is not satisfied which distinguishes it as a non-degenerative problem. Therefore, the exponential diffusivity allows propagation of the disturbances with infinite speed as in the classical diffusion equation.

Therefore, in order to clarify what type of diffusion we analyze let us look at Fig. 1 presenting the function $D/D_0 = \exp(\beta u)$ (β is dimensionless as it follows from the previous point) in the range $1 \geq u \geq 0$. Since $u = 1$ at the penetrant-polymer interface, we have a *slow diffusion* since *the ratio D/D_0 reduces from* e^β at $x = 0$ *down to* 1 at the edge of the penetration front. To be precise, the behaviour is the same as in the case of power-law diffusivity $D/D_0 = u^\beta$, where *the diffusivity decreases with increase in the distance measured from the interface $x = 0$ in the depth of the penetrated medium because we have $1 \geq u \geq 0$ and exponents are greater than unity*. Moreover, we can see from Fig.1 that for high values of β both the power-law and the exponential functions tend to converge with acceptable accuracy for practical uses (the plot in the inset is just an inverted version).

As a support of the above comments we may approximate the exponential function as a series (six terms for the sake of simplicity)

$$exp(\beta u) \approx 1 + \beta u + \frac{1}{2}\beta^2 u^2 + \frac{1}{6}\beta^3 u^3 + \frac{1}{24}\beta^4 u^4 + \frac{1}{120}\beta^5 u^5 + O(u^6) \quad (11)$$

and therefore the diffusion coefficient could be considered as a superposition of normal (Gaussian) diffusion (the 1st terms in RHS of (11) and diffusivities of power-law type with exponents greater than unity (diffusion equations of such type are degenerate parabolic equations, as commented above). Then, the diffusion equation could be approximately presented as

$$\frac{\partial u}{\partial t} = D_0 \frac{\partial u}{\partial x}\left\{\left[1 + \beta u + \frac{1}{2}\beta^2 u^2 + \frac{1}{6}\beta^3 u^3 + \frac{1}{24}\beta^4 u^4 + \frac{1}{120}\beta^5 u^5\right]\frac{\partial u}{\partial x}\right\} \quad (12)$$

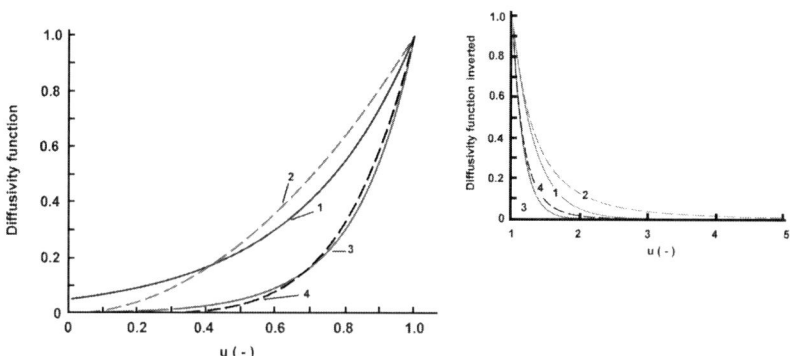

Figure 1. Comparisons of power-law $D/D_0 = u^\beta$ (lines 2 and 4) and exponential functional relationships $D/D_0 = exp(\beta - 1)u$ (lines 1 and 3) (similar to the analysis in [20]) of the diffusivity for different values of the exponent β, $0 < u < 1$; Lines 1 and 2 $\beta = 3$;Lines 3 and 4 $\beta = 6$. Inset : Inverted version of the dimensionless diffusivity showing that for decreasing in depth concentration and $\beta > 1$ the diffusion coefficient should decrease, that is we have a *slow diffusion*.

Taking into account that *these functional relationships come from experimental data fitting* [7, 9, 10, 11, 12, 16, 19, 25, 26, 27, 28, 29, 32, 33, 34, 35] but *not from theoretical concepts* (see the next section 1.4), we may see that the problem of the finite and infinite speeds of the solutions is a purely mathematical by nature. From this point of view this chapter assumes that irrespective of the type of the diffusion coefficient (its functional dependence of the concentration) the solutions (wetting front in case of porous media infiltration) propagate with finite speeds *since infinite speeds of the parabolic models are unphysical.* Moreover, we expect that at high values of β the solutions will exhibit strongly convex forms of sharp moving waves such as the solutions when the diffusivity is of a power-law type [23].

1.4. The Ranges of the Rate-Factor β in Real Physical Situations

Before starting further analyses and developing solutions we have to clarify what type of diffusion processes are modelled by the governing equations (5) or (6). The question, beyond the introductory comments in the lasts part of section 1.1, addresses the value of the coefficient in the exponential function. In cases when liquids (or gases) penetrate polymers the plastifization coefficient β (hereafter will use this symbol) varies in the range from 0.001 to 1 [9] for sorption of CO_2 in wood-fibre/polystyrene composites and from 0.2 to 2 [29] for diffusion in ethylene-propylene rubber, or $\beta = 4.8$ for sorption of heptane in low-density polyethylene [30]. In the numerical simulations of [38] the range $1 < \beta < 5$ was used, while Zhou et al. [31] have performed numerical simulations with β in the range $-0.5 < \beta < 0.5$.

In soil infiltration experiments and moisture diffusion in building materials the value of β attains values greater than 1, for example : $\beta = 3$ [32], $\beta = 8$ [20, 33] and $\beta = 20$ [20].

Therefore, in order to clarify what type of diffusion we analyze and model the plots in Figure 1 present the function $D/D_0 = \exp(\beta u)$ in the range $1 \geq u \geq 0$. Since $u = 1$ at the penetrant-polymer interface, we have a slow diffusion since the ratio D/D_0 reduces from e^β at $x = 0$ down to 1 at the edge of the penetration front. The behaviour is the same as in the case of power-law diffusivity $D/D_0 = u^\beta$, where the diffusivity decreases with increase in the distance measured from the interface $x = 0$ in the depth of the penetrated medium.

1.5. Existing Solution Approaches: A Short Overview

1.5.1. Boltzmann Similarity Transform Approach

Budd and Stockie [20] have studied the case when $D(0) \propto 1$ but $D(u) = O(1)$ for u away from zero, thus creating a nearly-degenerate problem with the idea that the solution will exhibit a steep front propagating with a finite speed.

The earlier idea to solve (2) is based on the Boltzmann transformation $\eta = x/(2\sqrt{D_0 t})$ [25, 38] to describe the penetration into semi-infinite media and this approach results in ODE, namely:

$$\frac{dC}{d\eta} = -\frac{1}{2\eta}\frac{d}{d\eta}\left(e^{\beta C}\frac{dC}{d\eta}\right), \quad C(\eta = 0) = 1, \quad C(\eta \to \infty) \to 0 \quad (13)$$

Otherwise, by defining the variable $\phi = e^{\beta C}$ [38] the exponential function can be eliminated which yields

$$\frac{d^2\phi}{d\eta^2} = -2\frac{\eta}{\phi}\frac{d\phi}{d\eta}, \quad \phi(\eta = 0) = 1, \quad \phi(\eta \to \infty) \to 0 \quad (14)$$

The numerical solution of (14) using Maple (the Rosenbrock scheme) is shown in Figure 2. The initial condition $d\varphi/d\eta=\text{-g}$ (determining the slope of the plot at $\eta \to 0$) was taken in accordance with the technique used by Budd and Stockie [20].

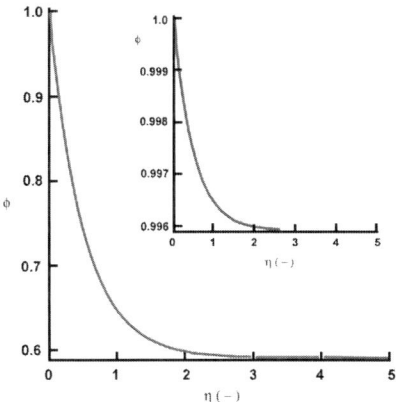

Figure 2. Numerical solution of equation (14) at $g = -0.818$ ($\beta \approx 0.2$); Inset: Numerical solution at $g = -0.00818$ ($\beta \approx 4.8$).

The solution with respect to ϕ can be converted to a solution to C by integration of

$$\frac{dC}{d\eta} = \frac{1}{\beta}\frac{1}{\phi}\frac{d\phi}{d\eta} \quad (15)$$

Numerical solutions of (13), (14) and (15) by the finite-difference method are discussed by Budd and Stokie [20] and by Riek et al.[38] (see comparisons to the Riek's solutions in section 5.1). Solutions with in initial diffusion equation transformation by the Boltzmann variable was employed in other approximate solutions [35, 55, 37] too.

Alternatively, defining the similarity variable as $\eta = \frac{x}{\sqrt{D_0 t e^\beta}}$ it is possible to transform the diffusion equation (5) into an ordinary differential equation

$$\Theta \frac{d^2\Theta}{dt^2} = -\eta_{bs}\frac{d\Theta}{dt}, \quad \eta_{bs} = \frac{x}{\sqrt{2D_0 t e^{\beta u}}}, \quad 0 < \eta_{bs} < \infty \quad (16)$$

where

$$\bar{\beta} = \beta \Delta u = \beta \left(u_{(x=0)} - u_{(t=0)}\right) \quad (17)$$

and

$$\Theta(0) = 1, \quad \Theta(\infty) = \varepsilon := e^{-\bar{\beta}} \quad (18)$$

It is worth-noting, that when diffusion equation models water wetting of porous media then both β and $\bar{\beta}$ are significantly greater than unity and consequently the parameter ε defined in (18) satisfies the condition $0 < \varepsilon \ll 1$ [20]. The postulation $\varepsilon \neq 0$ differs from the assumption of *a sharp front solution* where *no diffusion occurs beyond the penetration front*, commented in the next points and used in the integral-balance solutions developed in this chapter.

1.5.2. Asymptotic Solutions

Babu's Solution

Babu, studying water wetting of soils with exponential diffusivities [39], assumed that beyond a given position x^*, that is, for $x > x^*$, there is no diffusion process, actually value of water concentration is assumed equal to the residual value $\varepsilon \ll 0$ (but not equal to zero) and the governing equation is

$$\Theta \frac{d^2\Theta}{dt^2} = -x\frac{d\Theta}{dt} \quad (19)$$

with initial and boundary conditions

$$\Theta(0) = 1, \quad \Theta(x^*) = \varepsilon, \quad \left(\frac{d\Theta}{dx}\right)_{x \to x^*} \to 0 \quad (20)$$

Here the front position x^* is unknown and should be determined as a part of the solution procedure. The last boundary condition means that no water flux is assumed beyond x^* (that is we should set $d\Theta/dx = 0$. The Babu's approach utilizes a series expansion in $\eta_b = (1-\varepsilon)/\bar{\beta}$ which leads to the following approximation of the wetting front

$$\frac{d^2x^*}{dt^2} = \sqrt{\eta_b}\left[1 + \frac{1}{3_b} + \left(\frac{17}{90} + \frac{\bar{\beta}}{8}\right)\eta_b^2 + ...\right] \tag{21}$$

As Budd and Stokie [20] mentioned if the asymptotically small value $\varepsilon = exp(-\beta)$ is neglected then, the front position can be approximated as

$$x^*_{(Babu)} \approx (\bar{\beta})^{-1/2} + \frac{11}{24}(\bar{\beta})^{-3/2} + O\left(\bar{\beta}^{-5/2}\right) \tag{22}$$

where the rate factor $\bar{\beta}$ is defined as $\bar{\beta} = \beta(C_i - C_0)$ [20, 39]. Hence, $\bar{\beta}$ is dimensionless when the dependent variable is the concentration C. However, this is equivalent to the factor β used here (see eq. (5)) when the scaled (dimensionless) concentration u (4) is used.

Budd and Stokie Solution

An approximation of the wetting from position, similar to that of Babu, was obtained in [20] (see (16)), namely

$$x^*_{(Budd-Stokie)} \approx \frac{e^{1-\bar{\beta}}}{\gamma} + \bar{\beta}^{-1/2} + \frac{3}{4}\bar{\beta}^{-3/2} + O\left(\bar{\beta}^{-5/2}\right) \tag{23}$$

where

$$\gamma \approx \bar{\beta}^{1/2} - \frac{1}{4}\bar{\beta}^{-1/2} - \left(\frac{0.0866}{2} + \frac{1}{32}\right)\bar{\beta}^{-3/2} \tag{24}$$

When the solution should be re-written in terms of the physical variables then the relationship $\bar{C} = 1 + (1/\bar{\beta})\log C$ and $\varepsilon = exp(-\bar{\beta})$ should be applied for the transformations (see eq. (15)) where the integration leads to the mentioned relationship), that is the integration of the solutions shown in Figure 2.

Parlange's Solution

Parlange [34, 40] solution is based on an iterative approach to the general diffusivity definition

$$D(C) = -\frac{dx}{dC}\int_{C_0}^{C} x(\alpha)d\alpha \tag{25}$$

based on the measurement of Bruce and Klute [41] when $x(C)$ is a function of the porous medium saturation.

Similar to Babu and Budd-Stokie solutions, the Parlange approximation for the wetting front is

$$x^*_{(Parlange)} \approx \left(\bar{\beta}^{-1/2} + \bar{\beta}^{-3/2}\right)\left(1 - 2e^{-\bar{\beta}}\right) \qquad (26)$$

with the assumption that $\Theta \equiv \varepsilon$ for $x > x^*$. with asymptotic behaviour as [20, 40]

$$x^*_{(Parlange)} \approx \bar{\beta}^{-1/2} + \bar{\beta}^{-3/2} \qquad (27)$$

since the second term in (27) approaches unity very fast.

Further, Parlange and Babu [42] compared their solutions techniques and confirmed that both approaches (i.e. the perturbation and iterative methods) yield identical results. These approximation are simulated by Maple and shown in Figure 3. As it may be expected they converge for large values of β, actually for $\beta > 4$ (see the main figure and the inset).

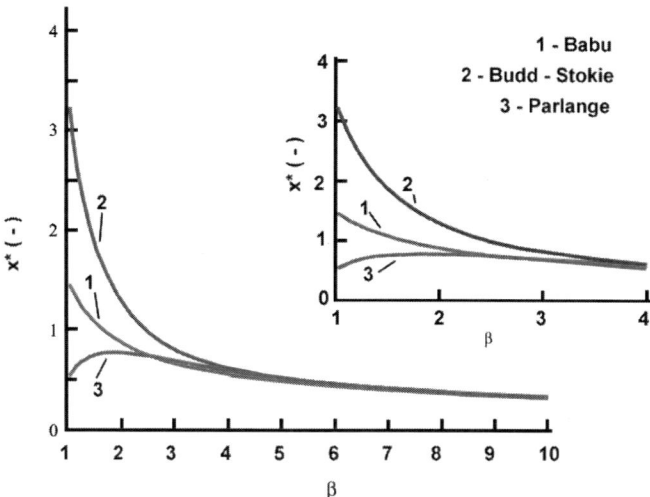

Figure 3. Comparison of the approximations of the dimensionless front x^* as a function of the rate factor β established by asymptotic solutions of Babu, Budd-Stokie and Parlange; Inset :The behaviour of these approximations for small β where they do not converge.

However, we have to mention an important issue, actually we address the solution concept, emerging in the solutions of Babu and Parlange, that is, *the finite depth of the penetration* (wetting) *front* (position) and we will comment this in the text point.

1.5.3. Moving Front Approaches

The finite depth of the penetration (wetting) front (position) is a physically motivated concept corresponding to the fact that the diffusion front (solution) should have a finite speed irrespective to the fact that the models is parabolic. The finite front position and its temporal evolution emerges also as a modelling concept in solvent diffusion into solid polymers [43]. From a mathematical point of view, the problem of penetrant diffusion into a polymer and the associated problem of diffusional solute release could be considered as a Stefan problem with phase-change boundary. In fact, such a boundary has to describe a sharp front separating the zone with absorbed solvent from the virgin polymer.

Since the problem at issue is truly parabolic, the idea of the moving front leads to the concept of the finite penetration depth. This concept has been used by Duda and Vrentas [43] to develop an approximate solution by the weighted residuals method [44] using, the zero and the first moment equations. The method of solution developed this chapter is from the family of the weighted residuals method and is restricted to the zero moment only with weighting function equal to 1 and commonly known as the *integral balance method* [45, 46, 47, 48]. In contrast to the concept of the finite penetration depth in the solutions of Babu and Budd and Stokie [20, 39] the integral-balance solution considers the front concentration flux equal to zero, i.e. the last boundary conditions in (20 are replaced by $\Theta(x^*) = \varepsilon = 0$ and $\left(\frac{d\Theta}{dx}\right)_{x \to x^*} = 0$, this making the solution non continuous at the front.

1.6. Motivation and Main Concept

The present chapter address approximate solution of the problem represented by (5) by the integral-balance method in two basic versions: Heat-balance Integral Method (HBIM) [45, 46] and the Double Integration Method (DIM) [49, 47, 48]. These methods have been commonly applied to diffusion problems (heat and mass) with constant diffusivities [46]. The recent solutions [22, 23] revealed that the approach led to successful results with power-law diffusivities. The solutions developed further is this article demonstrate how this method works in case of exponential diffusivity.

1.7. Organization of the Following Part of the Chapter

The following part of the chapter is organized as follows: Section 2 presents integral-balance solutions developed by HBIM and DIM. Section 3 focuses the attention on the optimization of the solution and the relation of the unspecified exponent of the assumed profile to the rate-factor β; Section 4 presents numerical experiments with the developed approximate solutions. Section 5 demonstrates comparisons of the developed integral balance solution to other approximate solution available in the literature.

2. APPROXIMATE INTEGRAL-BALANCE SOLUTION

The method suggests a finite penetration depth δ and replacing the condition $u \to 0$ as $x \to \infty$ by $u(\delta, t) = \partial u(\delta, t)/\partial x = 0$, thus defining a sharp front beyond which the medium is undisturbed. The concept of the penetration depth has been introduced in the approximate solutions of transient diffusion [45, 46] to compensate the principle deficiency of the parabolic equation: the infinite speed of propagation of the flux. The penetration depth concept and the integral-balance solution divide the solution in two steps: 1) A penetration step when the medium is considered as semi-infinite, and 2) Post-penetration step for the times beyond the moment when $\delta = L$. The most interesting case is the penetration period and we will focus our attention on it.

2.1. Integral-Balance Method

2.1.1. HBIM Integration Technique

Integrating (5) over the penetration depth (from 0 to δ) we get

$$\int_0^\delta \frac{\partial u}{\partial t} dx = \int_0^\delta \frac{\partial u}{\partial x}\left(D_0 e^{\beta u} \frac{\partial u}{\partial x}\right) dx \qquad (28)$$

$$\int_0^\delta \frac{\partial u}{\partial t} dx = D_0 e^{\beta u} \frac{\partial u}{\partial x}\bigg|_\delta - D_0 e^{\beta u} \frac{\partial u}{\partial x}\bigg|_0 \qquad (29)$$

and therefore

$$\frac{d}{dt}\int_0^\delta u\, dx = -D_0 e^{\beta u}\frac{\partial u}{\partial x}\bigg|_0 \qquad (30)$$

The relation (30) is the basic equation of the Heat-balance Integral method (HBIM) [45, 46]. Its physical meaning is that the time variation of the mass of diffusant penetrated into the medium (the integral in the LHS of (30) is controlled by the mass flux at the boundary $x = 0$ (the RHS of (30)).

Then replacing u by an assumed profile as function of the spatial co-ordinate (polynomial or exponential [45, 46] in (30) and integrating the result is an ordinary differential equation about δ. The main deficiency of HBIM is that the right-side of (30) contains the gradient $(\partial u/\partial x)_{x=0}$ which has to express through the assumed profile.

2.1.2. DIM Integration Technique

The first step of the DIM [23] is the integration from 0 to x and the result is

$$\int_0^x \frac{\partial u}{\partial t}\, dx = D_0 e^{\beta u}\frac{\partial u}{\partial x}\bigg|_x - D_0 e^{\beta u}\frac{\partial u}{\partial x}\bigg|_0 \qquad (31)$$

Further, representing the integration in left side of (30) as $\int_0^\delta f(\bullet)dx = \int_0^x f(\bullet)dx + \int_x^\delta f(\bullet)dx$ and then subtracting (31) from (30) we get an integral relation in the zone at the vicinity of the front δ, namely

$$\int_x^\delta \frac{\partial u}{\partial t}\, dx = -D_0 e^{\beta u}\frac{\partial u}{\partial x}\bigg|_x \qquad (32)$$

With $e^{\beta u}\frac{\partial u}{\partial x} = \frac{1}{\beta u}\frac{\partial e^{\beta u}}{\partial x}$, the integration of (32) from 0 to δ yields

$$\int_0^\delta \int_x^\delta \frac{\partial u}{\partial t}\, dx = D_0 \frac{1}{\beta u}e^{\beta u}\bigg|_{x=0} \qquad (33)$$

The expression (33) is the principle equation of the Double Integration Method (**DIM**) in case of exponential diffusivity.

2.2. Assumed Profile and Penetration Depths

The solutions envisage application of an assumed parabolic profile with unspecified exponent [46, 48, 50, 51], i.e.

$$u_a(x,t) = \left(1 - \frac{x}{\delta}\right)^n \quad (34)$$

With this profile we have $u_a(0,t) = 1$ and $u_a(\delta,t) = 0$, i.e. it satisfies the boundary conditions imposed by the finite penetration depth concept, for any value of n. The integral-balance method suggests replacement of $u(x,t)$ by $u_a(x,t)$ in in the integral relation (30) or (33) thus developing an equation about $\delta(t)$ as it is demonstrated next.

2.2.1. DIM

Further, replacing u by u_a in (33) we get an equation about $\delta(t)$, namely

$$\frac{1}{(n+1)(n+2)}\frac{d\delta^2}{dt} = D_0 \frac{1}{\beta}\left(e^\beta - 1\right) \quad (35)$$

$$\delta_{DIM} = \sqrt{D_0 t}\sqrt{(n+1)(n+2)}\sqrt{\frac{1}{\beta}\left(e^\beta - 1\right)} \quad (36)$$

Alternatively we get

$$\delta_{DIM} = \sqrt{D_0 e^\beta t}\sqrt{(n+1)(n+2)}\sqrt{\frac{1}{\beta}\left(1 - \frac{1}{e^\beta}\right)} \quad (37)$$

This solution excludes the case $\beta = 0$, but the limit of the last term of δ for $\beta \to 0$ (see the terms in the right-hand sides of (35) and (36)) is $\lim_{\beta \to 0}\left[\frac{1}{\beta}\left(e^\beta - 1\right)\right] = 1$ (because for small values of β we have $e^\beta \approx 1 + \beta$ and $\left(e^\beta - 1\right)/\beta \approx 1$). Therefore, from (37) one obtains $\lim_{\beta \to 0}\delta_{DIM} = \sqrt{D_0 t}\sqrt{(n+1)(n+2)}$ which is the classical DIM solution [49, 48]. However, we have to bear in mind, that *when diffusion in polymers is at issue*, the case with $\beta = 0$ means *missing free volumes and plasticizing action of the penetrant* and therefore, *absence of a diffusion process* [1, 7, 16, 17].

2.2.2. HBIM

From the HBIMs equation (30) and replacing u by u_a we get

$$\frac{1}{n+1}\frac{d\delta}{dt} = -D_0 e^\beta \left(-\frac{n}{\delta}\right) \tag{38}$$

$$\delta_{HBIM} = \sqrt{D_0 t}\sqrt{2n(n+1)e^\beta} \tag{39}$$

This integration technique does not exclude the case $\beta = 0$; for this limiting situation (39) reduces to the classical HBIM solution [45, 46].

2.3. The Approximate Profile

With the expressions about the penetration depth developed by the two methods of integration we get two different forms of the approximate solution

2.3.1. DIM Solution

$$u_{a(DIM)} = \left(1 - \frac{x}{\sqrt{D_0 t}F_1(n,\beta)}\right)^n = \left(1 - \frac{x}{\sqrt{D_0 t e^\beta}F_2(n,\beta)}\right)^n \tag{40}$$

$$F_1(n,\beta)_{DIM} = \sqrt{(n+1)(n+2)}\sqrt{\frac{1}{\beta}(e^\beta - 1)} \tag{41}$$

$$F_2(n,\beta)_{DIM} = \sqrt{(n+1)(n+2)}\sqrt{\frac{1}{\beta}\left(1 - \frac{1}{e^\beta}\right)} \tag{42}$$

The numerical factors $F_1(n,\beta)_{DIM}$ and $F_2(n,\beta)_{DIM}$ depend on the value of β which is pre-determined by the model (2) and the exponent n of the assumed profile.

2.3.2. HBIM Solution

$$u_{a(DIM)} = \left(1 - \frac{x}{\sqrt{D_0 t}F_1(n,\beta)_{HBIM}}\right)^n = \left(1 - \frac{x}{\sqrt{D_0 t e^\beta}F_2(n,\beta)_{HBIM}}\right)^n \tag{43}$$

$$u_{a(DIM)} = \left(1 - \frac{x}{\sqrt{D_0 t} F_1(n, \beta)_{HBIM}}\right)^n = \left(1 - \frac{x}{\sqrt{D_0 t e^\beta}, \; F_2(n, \beta)_{HBIM}}\right)^n \quad (44)$$

The numerical factor $F_1(n, \beta)_{HBIM}$ depends on the value of β which is pre-determined by the model (2) and the exponent n of the assumed profile, but $F_2(n, \beta)_{HBIM} = \sqrt{2n(n+1)}$ depends only on the exponent n of the assumed profile as in the classical **HBIM** solution.

2.4. Dimensionless Penetration Depths

The profiles (40) and (43) define directly two similarity variables:

$$\eta_B = \frac{x}{\sqrt{D_0 t}}, \quad \eta_\beta = \frac{x}{\sqrt{D_0 t e^\beta}} \quad (45)$$

The first one η_B is the Boltzmann similarity variable while η_β resembles the one (see (16) for the definition) used by Budd and Stockie in [20] in transformation of eq. (5) into eq. (16). Both $\sqrt{D_0 t}$ and $\sqrt{D_0 t e^\beta}$ can be considered as length scales, *since the semi-infinite medium problem has no its own length scale.* This allows to present dimensionless forms of the penetration depth (similar to the values of x^* in the asymptotic solutions) as

HBIM solution

$$\delta^*_{HBIM-1} = \frac{\delta_{HBIM}}{\sqrt{D_0 t}} = \sqrt{2n(n+1)e^\beta} \quad (46)$$

and

$$\delta^*_{HBIM-2} = \frac{\delta_{HBIM}}{\sqrt{D_0 t e^\beta}} = \sqrt{2n(n+1)} \quad (47)$$

DIM solution

$$\delta^*_{DIM-1} = \frac{\delta_{DIM}}{\sqrt{D_0 t}} = \sqrt{(n+1)(n+2)} \sqrt{\frac{1}{\beta}(e^\beta - 1)} \quad (48)$$

and

$$\delta^*_{DIM-2} = \frac{\delta_{DIM}}{\sqrt{D_0 t e^\beta}} = \sqrt{(n+1)(n+2)} \sqrt{\frac{1}{\beta}\left(1 - \frac{1}{e^\beta}\right)} \quad (49)$$

Hence, the dimensionless penetration depths depend only on the values of the exponent n and the rate-factor β similar to x^*. We will see in section 3.1.2 that n can be presented as a function of β only (see eq.(62)), thus making the dimensionless penetration depth dependent only on β. The comparative plots presented in Figure 3 reveal that the dimensionless penetration depth established by HBIM and DIM, when the length scale is $\sqrt{D_0 t e^\beta}$, matches to a greater extent the approximations developed by the asymptotic methods (precisely to that of Babu). This is obvious even for values of $\beta < 4$ where the Budd-Stokie's approximation provides extremely large values, while the Parlange's approximation decreases as $\beta \to 0$. Physically thinking, the larger penetration depth corresponds to the case when $\beta = 0$, so all these approximations are physically adequate. With increase in the value of β the penetration depth becomes shorter and slowly tends towards an almost constant value; actually we get physically indistinguishable dimensionless penetration depths (see the estimations of the limits above and the assessments in section 4.1).

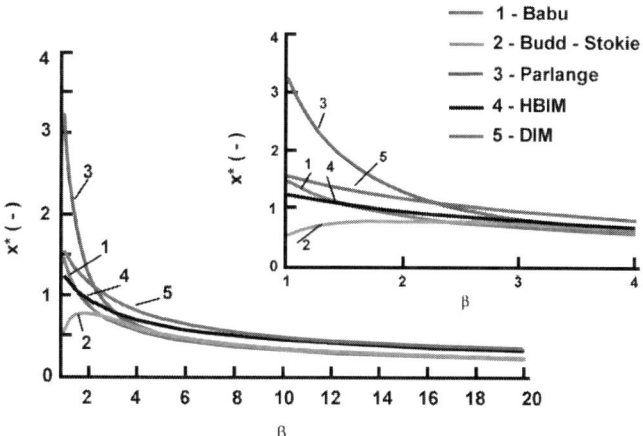

Figure 4. Comparison of the approximations of the dimensionless front x^* as a function of the rate factor β established by asymptotic solutions (Babu, Budd-Stokie and Parlange) and the dimensionless penetration depth determined by HBIM and DIM when the length scale is defined by $\sqrt{D_0 t e^\beta}$; Inset: Penetration depth for small values of β where they are distinguishable. For the HBIM and DIM approximation the relationship $n_\beta = 1/(1+\beta)$ is used (see section 3.1.2).

2.5. Front Time Evolution Scaling

The principle question arising from these solutions is : what is the time scaling of the penetration depths, that is of the solutions developed. The question raised is related to the fact that the power-law diffusivity $D(u) = u^m$, for $m > 1$ results in a reduced penetration depth, but the Gaussian behaviour of the front remains (Dirichlet problem, see in the sequel eq.(53)), despite the convex solution profile [23]. The plots in Figure 4 reveal Gaussian behaviours but with different behaviour numerical factors depending on n and β.

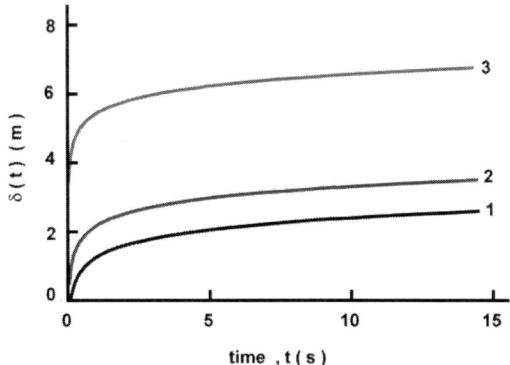

Figure 5. Time-scaling of the penetration depths. DIM solutions with different n and β. Line 1- Gaussian behaviour with $\beta = 0$ and $n = 2$, and $\delta \approx 3.464\sqrt{t}$; Line 2-DIM solution $\delta_{DIM-2} = 8.632\sqrt{t}$ with $n = 2$ and $\beta = 5$; Line 3- DIM solution $\delta_{DIM-1} = 229.13\sqrt{t}$ with $n = 2$ and $\beta = 0.5$; For the sake of simplicity it was assumed $D_0 = 1$.

However, the presentation depth could be expressed in a different way. Let us use, the series approximation (11). In this case the model equation (5) can be presented (with 4 terms of the series expansion only for the sake of simplicity) as

$$\frac{\partial u}{\partial t} \approx D_0 \left[\frac{\partial^2 u}{\partial x^2} + \frac{\beta}{2} \frac{\partial^2 u^2}{\partial x^2} + \frac{\beta^2}{6} \frac{\partial^2 u^3}{\partial x^2} \right] \qquad (50)$$

This presentation strongly reveals that we have a superposition of degenerate terms. To some extent this may explain why the solution profiles are convex (see further in the text of section 4) in contrast to concave profiles of the Gaussian model when only the fist term takes place, i.e for $\beta = 0$.

Then applying the DIM technique we get

$$\frac{1}{(n+1)(n+2)}\frac{d\delta^2}{dt} = D_0 t \left[1 + \frac{\beta}{2} + \frac{\beta^2}{6}\right](n+1)(n+2) \quad (51)$$

That is,

$$\delta = \sqrt{D_0 t} \sqrt{\left[1 + \frac{\beta}{2} + \frac{\beta^2}{6}\right](n+1)(n+2)} \quad (52)$$

When, only one degenerate diffusion coefficient exists, for instance $D(u) = D_0 u^m$, then the penetration depth scales as [23]

$$\delta_{degenerateonly} \equiv \sqrt{D_0 t} \sqrt{\frac{(n+1)(n+2)}{m+1}} \quad (53)$$

Hence, the increase in the value of m, reduces the penetration depth and the solution approaches a sharp wave behaviour (see section 4) (more detailed studies are available elsewhere [23])

3. OPTIMIZATION OF THE APPROXIMATE SOLUTION

The use of a parabolic profile with unspecified exponent gives us flexibility in defining the optimal approximate solution. Precisely, since the assumed profile satisfies the boundary condition for any value of n then defining the error of approximation as a function of it we may define the optimal exponent and therefore the optimal approximate profile.

3.1. Residual Function

The residual function $R(x, t, \beta, n)$ can be defined directly from (5) by replacement of u by u_a developed by either HBIM of DIM solution, namely:

$$R(x, t, \beta, n) = \frac{\partial u_a}{\partial t} - D_0 e^{\beta u_a}\left[\beta\left(\frac{\partial u_a}{\partial x}\right)^2 + \frac{\partial^2 u_a}{\partial x^2}\right] \quad (54)$$

The condition $R(x, t, \beta, n) = 0$ is obeyed only by the exact solution $u(x, t, \beta)$, if it exists. Otherwise, when $u \to u_a$ we should have $R(x, t, \beta, n) \to 0$. Since this is impossible, the strategy to define the optimal exponent addresses a minimization of $R(x, t, \beta, n) \to min$ with respect to n as independent variable.

3.1.1. Residual Function in Terms of the Similarity Variable

In order to minimize the number of the variable in $R(x, t, \beta, n)$ we will express the approximate profile through the similarity variable η as $u_a = (1 - \eta/F_1)^n$ (see eqs. (40) and (42)).

Further with

$$\frac{\partial \eta_B}{\partial t} = \left(-\frac{x}{F_1}\right) t^{-3/2}, \quad \frac{\partial \eta_B}{\partial x} = -\frac{1}{\sqrt{D_0 t}} \quad (55)$$

and

$$\frac{\partial u_a}{\partial t} = -n\left(1 - \frac{\eta_{NB}}{F_1}\right)^{n-1} \frac{\partial \eta}{\partial t}, \quad \frac{\partial u_a}{\partial x} = -n\left(1 - \frac{\eta_B}{F_1}\right)^{n-1} \frac{\partial \eta_B}{\partial x} \quad (56)$$

we have

$$R(\eta, \beta, n) = -\frac{xt^{-3/2}}{\sqrt{D_0}} n \left(1 - \frac{\eta}{F_1}\right)^{n-1} -$$

$$-\frac{D_0 e^{\beta u_a}}{F_1^2 (D_0 t)} \left(1 - \frac{\eta}{F_1}\right)^{n-2} \left[\beta n^2 \left(1 - \frac{\eta}{F_1}\right)^n + n(n-1)\right] \quad (57)$$

More, with $0 \leq X = \eta/F_1 \leq 1$ we transform the moving boundary domain in to a fixed boundary range for X, which allows the residual function to be presented as

$$R(X, \beta, n, t) = \frac{1}{t} R_0(X, \beta, n) \quad (58)$$

where R_0 is the time-independent term (in the large brackets) of (58), i.e.

$$R_0(X, n\beta) = XF_1 n(1 - X)^{n-1} -$$
$$\exp\left[\beta(1-X)^n\right] \frac{1}{F_1^2} (1-X)^{n-2} \left[\beta n^2 (1-X)^n + n(n+1)\right] \quad (59)$$

taking into account that in the evaluation of $R(\eta, \beta, n)$ the emerging product $\delta (d\delta/dt) = F_1^2$ is time-independent (that can be easily checked).

Since in the fixed domain $0 \leq X \leq 1$ the residual function is decaying in time, then the minimum of $R(\eta, \beta, n)$ depends on n if the value of β is stipulated. More correct approach is to minimize the mean-squared value of $R(\eta, \beta, n)$ over the penetration layer, that is in terms of X we should have

$$E(n, \beta) = \int_0^1 [R_0(X, n, \beta)]^2 \, dX \to min \quad (60)$$

3.1.2. Residual Function at the Boundaries of the Penetration Layer

From (57) it follows that for $n = 0$ we have $R(\eta, \beta, n) = 0$ but this solution is unphysical. Then, setting the condition $R(0, \beta, n) \geq 0$ we get

$$R(0, \beta, n) = -\frac{D_0 e^{\beta u_a}}{F_1^2 (D_0 t)} [\beta n + (n-1)] \geq 0 \quad (61)$$

Therefore, at the boundary $x = 0$ the residual function satisfies the condition $R(0, \beta, n) \geq 0$ for

$$n \geq \frac{1}{1+\beta} \quad (62)$$

We have to mention, for instance, that in case of only one degenerate term in the right-hand-side of the diffusion model (i.e. $\frac{\partial}{\partial x}\left[D_0 u^m \frac{\partial u}{\partial x}\right]$ the optimal exponent, in general follows the rule $n = \frac{1}{m}$; and for large m, the rule $n = \frac{1}{m+1}$ is valid, too [23].

Further, for the limit $\eta \to F_1$, that is $u_a \to 0$ in the zone near the front, we have from (57) that $R(F_1, \beta, n) = 0$ for any n (see the last term of eq.(57)) the condition is $n \leq 1$. In fact, the condition at $\eta \to F_1$ is in agreement with that at $x = 0$. Therefore, the exponent of the approximate profile defined in this specific case resembles n_{PL} corresponding to approximate integral-balance solution of a degenerate diffusion equation with power-law diffusivity ($D = D_0 u^m$) [23]; at $x = 0$ the condition is $n_{PL} = 1/(1+m)$, for $m > 1$ [23]. In general, the minimization of $R(\eta, \beta, n)$ at the boundaries of the penetration depth imposes the restriction $1 \geq n \geq 1/(1+\beta)$ which indicates a quasi-degenerate behaviour of the diffusion equation leading to almost infinite gradient of the profile when $x \to \delta$, that is, $\eta \to F_1$ and increase in the parameter β.

3.1.3. Optimal Exponents of the Solutions

The minimization of the squared residual function (see eq. (60)) yields the optimal exponents of the profiles. Here we skip the cumbersome calculations since the technique is known [23, 48, 51] and can be applied for any case with $\beta > 0$. The plot in Figure 6 shows the optimal exponents (the points) for some moderate values of β and the limiting line $1/(1+m)$ defined by (62).

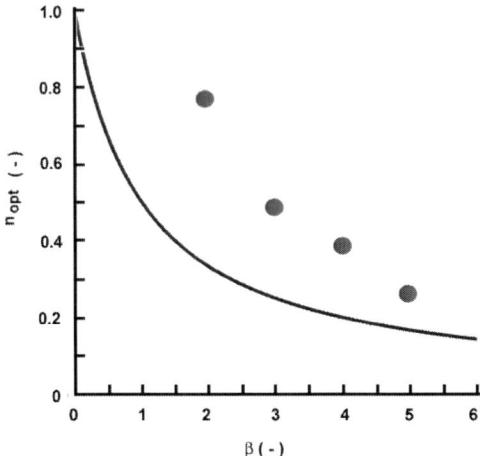

Figure 6. Optimal exponents for cases with moderate values of the rate-factor β and DIM solutions. Note: $n_{\beta=2} \approx 0.769$, $n_{\beta=3} \approx 0.487$, $n_{\beta=4} \approx 0.384$ and $n_{\beta=5} \approx 0.277$. The solid line represents the limit defined by (62).

4. NUMERICAL EXPERIMENTS WITH THE APPROXIMATE SOLUTION

4.1. Dimensionless Penetration Depth: Some Limits

Now with eq. (62)

$$n_\beta = \frac{1}{1+\beta} \qquad (63)$$

we may see some limits of this approximation, namely
For small values of β, i.e. when $0 < \beta < 1$, we have

$$1 < \frac{1}{1+\beta} < \frac{1}{2} \qquad (64)$$

For moderate values of β, i.e. when $\beta > 1$ we have

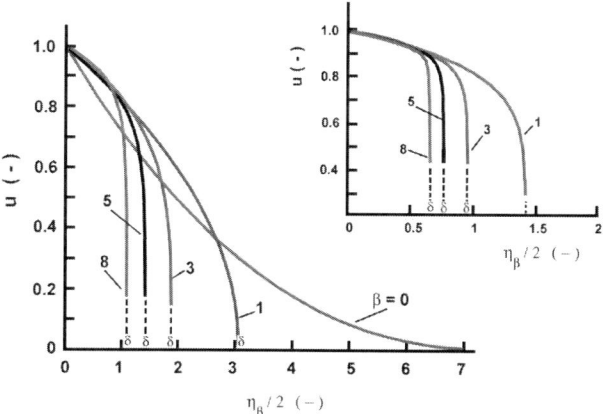

Figure 7. Approximate concentration profiles developed by DIM with $n = 1/(1+\beta)$ and presented as functions of the similarity variable η_B (i.e. functions of the common Boltzmann variable $x/2\sqrt{D_0 t}$ and moderate values of the rate constant β within the range $0 < \beta < 10$. Inset: Short distance profiles; Note: the dashed lines crossing the abscissa and the symbols δ denote the fronts of the solutions (penetration depths).

$$0 < n_\beta = \frac{1}{1+\beta} < \frac{1}{2} \qquad (65)$$

while for large β, when $\beta \gg 1$

$$0 < n_\beta = \frac{1}{1+\beta} \ll 1 \qquad (66)$$

Now, the dimensionless penetration depths as approximate functions depending only on β and they limits are

HBIM Solution

$$\delta^*_{HBIM-1} = \frac{\delta_{HBIM}}{\sqrt{D_0 t}} = \sqrt{2\frac{1}{\beta}(1+\frac{1}{1+\beta})e^\beta} \qquad (67)$$

For small values of $\beta \to 1$ we get $\delta^*_{HBIM-1} \to \sqrt{3e} \approx 2.8557$ while for $\beta \gg 1$ we have $\delta^*_{HBIM-1} \to \sqrt{2}\sqrt{\frac{e}{\beta}} \approx 2.3316$

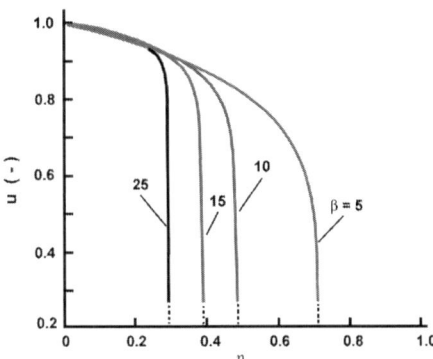

Figure 8. Approximate concentration profiles developed by DIM and $n = 1/(1+\beta)$, presented as functions of the similarity variable $\eta_B/2$ for large values of the rate constant β. The solutions are approaching step fronts (steeper that those in Figure 6) as functions of the similarity variable $\eta_B = x/\sqrt{D_0 t}$ generated naturally by the integral-balance solution.

$$\delta^*_{HBIM-2} = \frac{\delta_{HBIM}}{\sqrt{D_0 t e^\beta}} = \sqrt{2\frac{1}{1+\beta}(1+\frac{1}{1+\beta})} \qquad (68)$$

For $0 < \beta << 1$ we have $\delta^*_{HBIM-2} \to \sqrt{3/2} \approx 1.2248$. In the case of large $\beta >> 1$ the approximation limits, for example, are $\delta^*_{HBIM-2} \to 0.099773$ when $\beta \to 20$, and $\delta^*_{HBIM-2} \to 0.19835$ when $\beta \to 10$. Hence, with increase in the value of β the penetration depth decreases (typical behaviour for the degenerate diffusion) [22, 23].

DIM Solution
In a similar way we may establish that
for $0 < \beta << 1$

$$\delta^*_{DIM-1} = \frac{\delta_{DIM}}{\sqrt{D_0 t}} = \sqrt{(1+\frac{1}{1+\beta})(2+\frac{1}{1+\beta})}\sqrt{\frac{1}{\beta}(e^\beta - 1)} \qquad (69)$$

Then, for $\beta \to 1$ the limit is $\delta^*_{DIM-1} \to \frac{15}{4}(e-1) \approx 6.4440$.

and

$$\delta^*_{DIM-2} = \frac{\delta_{DIM}}{\sqrt{D_0 t e^\beta}} = \sqrt{(1+\frac{1}{1+\beta})(2+\frac{1}{1+\beta})}\sqrt{\frac{1}{\beta}\left(1-\frac{1}{e^\beta}\right)} \quad (70)$$

For $\beta \to 1$ (but $\beta > 1$) we have $\delta^*_{DIM-2} \to 2.3704$

Visually all these limits are illustrated by the plots in Figure 3 and Figure 4 commented earlier.

5. COMPARISON OF THE APPROXIMATE SOLUTION TO OTHER SOLUTIONS OF THE PROBLEM

It is hard to compare the integral-balance approximate solutions directly to some already solved examples due the variety of methods applied and the forms of results presentation. Because of that the following example are focused on cases where concentration profiles are presented as functions of the Boltzmann similarity variable $\eta_B = x/\sqrt{Dt}$. This is not easy task due to many different approaches in conversions of the initial modelling equation into non-linear ordinary equations and unnecessary nondimensalizations. Last but not least, the use of unconventional (non SI) units requires in some cases conversion (scaling) factors to be introduced.

5.1. Solution of Riek et al.[38]

The solutions of Riek et al. [38] (with respect to eqs.(13) , (14) and (15)) where mentioned in section 1.5.1). This study deals with moderate values of β in the range $0 \leq \beta \leq 5$. The plots in Figure 9 compare the DIM solutions (with the optimal exponents shown in Figure 8) and the Riek's numerical. It is obvious that the main discrepancy of both solutions appears close to the front of the profiles; this is a common characteristic of the integral balance method defining a sharp front in contrast to analytical and semi-analytical methods commented at the beginning.

5.2. Solution of Lockington et al.[52]

The solution of Lockington et al. [52] belongs to the family of solutions where the initial diffusion equation is transformed into an ordinary equation using the

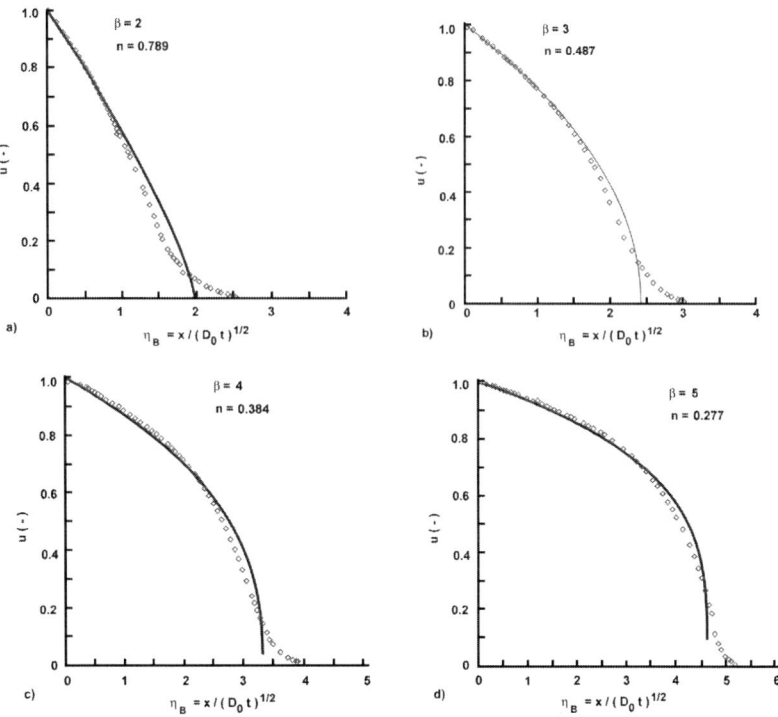

Figure 9. Comparison of DIM solutions (solid lines) with optimal exponents (see Figure 8) and numerical solution developed by Riek et al. [38] (points) by application of the Boltzmann similarity variable $\eta_B = x/\left(2\sqrt{Dt}\right)$.

Boltzmann variable in the form $\phi = x/\sqrt{t}$ (we use here the original notations and further the expressions will be converted to the notations used in this chapter) as it was shown in section 1.5.1. The ordinary equation obtained in this way is

$$-\frac{1}{2}\phi\frac{d\theta}{d\phi} = \frac{d}{d\phi}\left(D\frac{d\theta}{d\phi}\right) \qquad (71)$$

with initial and boundary conditions: $\theta(\phi = 0) = 1$, $\theta(\phi \to \infty) = 0$.

Here θ is the dimensionless concentration $\theta = \frac{C-C_i}{C_s-C_i}$ (where $C_s = C(t \to \infty)$-final concentration, $C_i = C(x, t = 0)$-initial concentration) and corresponds to the present (used in this chapter) dimensionless variable u.

Moreover, this solution uses the sorptivity function [53] defined as

$$S = (C_s - C_i) \int_0^1 \phi d\theta \tag{72}$$

An approximate solution [54] of this problem is

$$2 \int_0^1 \frac{D(\theta)}{a} da = S_{red}\phi + \frac{A}{2}\phi^2 \tag{73}$$

where the scaled sorptivity is defined as

$$S_{red} = \frac{S}{C_s - C_i} \cong \left[\int_0^1 (1+\theta) D(\theta) d\theta\right]^{\frac{1}{2}}, \quad A = 2 - \frac{S_{red}^2}{\int_0^1 D(\theta) d\theta} \tag{74}$$

With $D(\theta) = D_0 \exp(\beta\theta)$ equation (73) takes the form

$$2D_0 [E_i(\beta) - E_i(\beta\theta)] = S_{red}\phi + \frac{A}{2}\phi^2 \tag{75}$$

where $E_i(t) = -\int_{-z}^{\infty} \frac{e^{-t}}{t} dt$ is the exponential integral that should be approximated by a series or its tabulated values have to be used.

Then,

$$S_{red}^2 = D_0 \left[e^{\beta}\left(\frac{2}{\beta} - \frac{1}{\beta^2}\right) - \frac{1}{\beta} + \frac{1}{\beta^2}\right], \quad A = \frac{\frac{e^{\beta}}{\beta} - 1 - \frac{1}{\beta}}{e^{\beta} - 1} \tag{76}$$

The solution performed by Lockington et al. [52] with $D = 0.49e^{6.55\theta}$ ($D_0 = 0.49 \ [mm^2/min]$) is shown by points in Figure 10, just replacing the variable θ by u. Moreover, in order to make this solution comparable to the integral-balance solutions, the argument should be neither ϕ nor η_B but $z = \eta_B \sqrt{D_0}$. The reason of this scaling is the fact that the variable $\phi = x/\sqrt{t}$ has a dimension $[m/s^{1/2}]$ while the true Boltzmann variable $\eta_B = x/\sqrt{D_0 t}$ is dimensionless; Recall, $\sqrt{D_0 t}$ is a length scale in diffusion problems taking place in a semi-infinite medium (see the initial and boundary conditions related to (71)).

Additional problems that should be mentioned are: the diffusivity is presented in $[mm^2/\min]$, while ϕ is in $[cm/\sqrt{hour}\,]$. All these non SI units were transformed to $[m^2/s\,]$ and $[m/\sqrt{s}\,]$, correspondingly, in order to compare the solutions to the results obtained with the DIM technique.

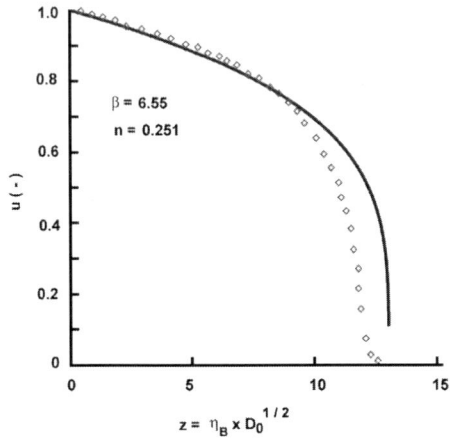

Figure 10. Comparison of DIM solutions (solid line) with optimal exponent $n \approx 0.251$ and solution developed by Lockington et al. [52] (points) by application of the Boltzmann similarity variable $\eta_B = x/\left(2\sqrt{D_0 t}\right)$ and $\beta = 6.55$. Note: In this case, the Boltzmann similarity variable was used in the form $x/\sqrt{t}$ (see the explanations in the text), thus the abscissa is in $\eta_B \times \sqrt{D_0}$, where $D_0 = 0.49$ $[mm^2/min]$.

The plots in Figure 10 reveal that both solutions are too close in the range where the concentration gradient is not so high, but the discrepancies between them are mainly near the steep fronts, where the concentration gradients may approach infinite values (see the almost vertical line of the DIM solution). However, as mentioned in the previous point, such behavior is typical when comparing integral-balance solutions and any other approximate analytical solutions of diffusion equations (where finite penetration depth is not considered).

5.3. Solution of Tzimopoulos et al. [55]

The solution developed by Tzimopoulos et al. [55] about horizontal infiltration of water in soils follows almost the same idea to transform the initial diffusion equation with diffusivity

$$D(C) = D_r e^{\beta_0 (C - C_i)} \tag{77}$$

where the initial concentration (water content) C_i is taken into account.

With the nondimensalization

$$\theta = \frac{C - C_i}{C_s - C_i}, \quad X = \frac{x}{L}, \quad \tau = D_r \frac{t}{L^2}, \quad D = \frac{D}{D_r} e^{\beta_0 \theta} \tag{78}$$

the model equation takes the form

$$\frac{\partial \theta}{\partial \tau} = \frac{\partial}{\partial X} \left(D_1(\theta) \frac{\partial \theta}{\partial X} \right) \tag{79}$$

with initial and boundary conditions

$$\theta(X \geq 0, \tau = 0) = 0, \quad \theta(X = 0, \tau > 0) = 1, \quad \theta(X \to \infty, \tau = 0) = 0 \tag{80}$$

where L is a characteristic length scale.

Moreover D_r in the last expression of (78) corresponds to the initial diffusivity D_0 used in this chapter.

We have to stress the attention on the fact that the introduction of characteristic length scale L contradicts the third boundary condition in (80) which defines diffusion in a semi-infinite medium; see the formulation of a such problem by eqs. (8)-(10). However, from general point of view, this will affect the final solution with appearance of unnecessary coefficients. Then, introducing the Boltzmann similarity variable as $\varphi^* = X/\sqrt{\tau}$ (in original notations) the models solved is

$$\frac{d}{d\varphi^*} \left(D_1 \frac{d\theta}{d\varphi^*} \right) + \frac{\varphi^*}{2} \frac{d\theta}{d\varphi^*} = 0, \quad \theta(\varphi^* \to \infty) = 0, \quad \theta(\varphi^* = 0) = 1 \tag{81}$$

with sorptivity in the form $S^* = \int_0^1 \varphi^* d\theta$ and using the solution of Philip and Knight [56]. Further, with a re-formulations of the diffusivity as $D_1(\theta) = e^{\lambda_1 \theta}$,

$D_1 = D/D_r$ (this is the same scaling as eq. (17) used by Budd and Stokie [20]) the final equation about φ^* is

$$\varphi^* = \frac{1}{S^*}\left[\frac{1}{\lambda_1}\left(e^{\lambda_1} - e^{\lambda_1\theta}\right) + E_i(\lambda_1) - E_i(\lambda_1\theta)\right] \qquad (82)$$

which resembles equation (75) of Lockington et al. [52]; nothing strange in this, since in spite of the initial nondimensalization procedure, the transformation of the initial model into an ordinary differential equation is one and the same (by the Boltzmann similarity variable).

The initial nondimensalization, commented above, introduces only additional coefficients, for example, the rate-factor of diffusivity function used to solve the final model is related to the initial formulation as $\lambda_1 = \beta_0(\theta_s - \theta_r)$. This needs re-calculation of the coefficients when this solution and the DIM solution should be compared. For this purpose all data summarized in Table 3 of the original work [55] are used. In addition, the unnecessary scaling the diffusivity function and the rate factor results in scaling coefficient γ (see the legends in Figure 11), so that the argument of the solution is not

$$\eta_B = x/\sqrt{D_0 t}$$

but it should be

$$\eta_\gamma = \eta_B/\gamma.$$

Moreover, the results presented in Figure 2 of [55] are in the so-called "reduced" (transformed) concentration profiles (the scale factors denoted as s are shown in the legends of Figure 11). The results in Figure 11 are presented replacing λ_1 by β in order to keep common notations in this chapter and consistency of the presentations.

The comparison of the DIM solutions in Figure 11 shows very good agreements especially in the regions of the steep fronts. This is in contrast to the all previously examples, where the discrepancy commonly appears in the area of the front.

5.4. Briefs on the Comparisons of Solutions

The example used in this section to compare the DIM solution of the problem at issue straightforwardly reveal the applicability of the integral-balance techniques to this non-linear problem. The main problem when comparing results

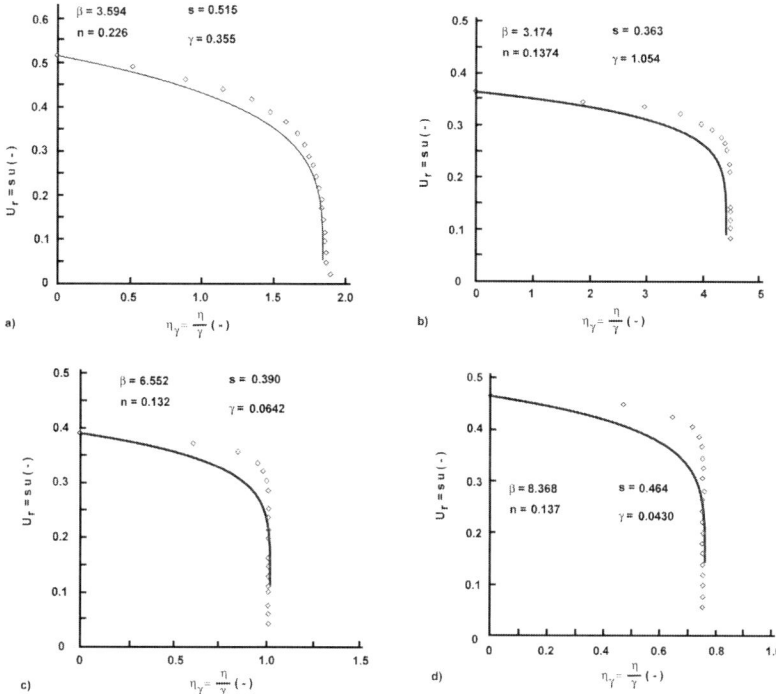

Figure 11. Comparison of DIM solutions (solid lines) with optimal exponents and approximate analytical solution of Tzimopoulos et al. [55] (points) about horizontal filtration in soils (by application of the Boltzmann similarity variable $\eta_B = X/(\sqrt{\tau})$), where: $\tau = D_r(t/L^2)$, $X = x/L$, L-characteristic length scale and $D_r = D_0$ in the present notation. The plots are in the so-called "reduced form" (the scale factors s and γ are inserted in the figure legends: For more details see the related text and Table 3 of the original work). Subfigures: a) Plots about *Hayden sandy loam* (Fig 2B in the original work; b) Plots about *Manawatu fine sand loam* (Figure 2C in the original work); c) Plots about *Adelanto loam* (Figure 2D in the original work); d) Plots about *Pine silty clay* (Figure 2I in the original work).

taken from different sources is the voluntary application of the Boltzmann similarity variable by changing its nature and introducing unnecessary scaling parameters. Actually, all solutions are on diffusion in semi-infinite medium, thus the initial step of Tzimopoulos et al. [55] makes the problem more complicated. Furthermore, the appearance of the exponential integral $E_i(t)$ as element of the solution procedure needs its approximation as series (as it was done in [55], but for small values of λ_1) or using programs as MATLAB where the accuracy of approximation is not well defined). Despite this, all commented solutions, and the integral-balance solutions used here, are approximate in nature since the model at issue has no exact solution. Therefore, the good agreement between them confirm the applicability of the integral-balance approach in this case.

CONCLUSION

This chapter presented integral-balance solutions of the non-linear diffusion equation with exponentially concentration dependent diffusivity. Precisely, the problem solved corresponds to cases with exponentially increased diffusivity which, to some extent, approaches the case with a power-law dependence on concentration (the so-called degenerate diffusion equation), even though at $C = 0$ the diffusivity function goes to 1 rather to zero as in the degenerate models. This was especially demonstrated in this chapter explaining why the convex profiles of the integral balance solution with exponential diffusivity resemble the solutions obtained with power-law diffusion function.

Furthermore, the integral balance solutions use the idea of a finite penetration depth which from different point of view is also used in the asymptotic solutions where it is expressed as a function of the parameter β. The comparison of the dimensionless penetration depths reveals that for large $\beta > 2$ the front approximated by the asymptotic methods and the integral-balance penetration depth have almost equal behaviours and there is no significant differences between them.

The comparative numerical experiments with known approximate solutions indicate acceptable convergences between them and the developed concentration profiles by the DIM technique (the solutions of Riek et al. [38] and Lockington et al. [52]). The common case is when the convergence appears for low values of the Boltzmann similarity variable, while the discrepancies appear at the solution front, except in the case solved by Tzimopoulos et al. [55] where the situation is just the opposite.

Last, we may say that the integral-balance approach is quite versatile with respect to non-linear diffusion problems with degenerate or non-degenerate diffusion functions (for this case see the commented reference sources in this chapter) thus allowing simple and straightforward solutions. Moreover, its simple techniques (HBIM or DIM) to define the functional relationship of the penetration depth allow definitely to determine the similarity variable (of Boltzmann or non-Boltzmann type). This step allows consequently correct applications to either the classical approach transforming the basic model to ordinary differential equation or to asymptotic solutions.

REFERENCES

[1] Neogi P. *Diffusion in Polymers*. Marcel Dekker, New York, 1996.

[2] Huang S.J. and Durning C.J., Nonlinear viscoelastic diffusion in concentrated polyester-ethylbenzene solutions, *J. Polymer Sci., B: Polym. Phys.* **35**, 2103–2119 (1997).

[3] Huang J.C., Liou H. and Liu Y.I., Diffusion in polymers with concentration dependent diffusivity, *Intern. J. Polymeric Mater.* **49**, 15–24 (2001).

[4] Simpson W. T., Determination and use of moisture diffusion coefficients to characterize drying of northern red oak (*Quercus rubra*), *Wood and Fiber Sci.* **27**, 409–420 (1993).

[5] Simpson W.T. and Liu J.Y., An optimization technique to determine red oak surface and internal moisture transfer coefficients during drying, *Wood and Fiber Sci.* **29**, 312–318 (1997).

[6] Leech C., Lockington D. and Dux P., Unsaturated diffusivity functions for concrete derived from NMR images, *Materiaux et Constructiions* **36**, 4132–18 (2003).

[7] Camera-Roda G. and Sarti G.C., Mass transport with relation in polymers, *AIChE J* **36**, 851860 (1990).

[8] Lin S.H., Drug release into external absorber: Concentration-dependent diffusivity, *AIChE J.* bf 55, 548–553 (2009).

[9] Doroudani S., Chaffey C.E. and Kortdchot M.T., Sorption and diffusion of carbon dioxide in wood-fiber/polystyrene composites, *J. Polym Sci.* **40**, 723–735. (2002).

[10] Joannes S., Maze L. and Bunsell A.R., A simple method for modeling of concentration-dependent water sorption in reinforced polymeric materials, *Composites: part B* **57**, 219–227 (2014).

[11] Gardner W.R. and Mayhugh M.S., Solutions and tests of the diffusion equation for movement of water in soils, *Soil. Sci/ Soc, Am. Proc.* **22**, 197–201 (1958)..

[12] McBride J.F. and Horton R., An empirical function to describe measured water distribution from horizontal infiltration experiments, *Water Resour. Res.* **21**, 1539–1544 (1985).

[13] Vrentas J.S. and Duda J.L., In : *Encyclopedia of polymer science and Engineering*, N. Bikales and J. Kroschewitz, Eds., John Wiley, New York, 1986.

[14] Vrentas J.S. and Duda J.L., Diffusion in polymersolvent systems. III. Construction of Deborah number diagrams, *J. Polym.Sci. Polym.Phys. Ed.* **15**, 441453 (1977).

[15] Durning C.J. and Tabor M., Mutual diffusion in concentrated polymer solutions under a small driving force, *Macromolecules* **19**, 2220–2232 (1986).

[16] Long R.B., Liquid penetration through plastic films, *Ind. Eng. Chem. Fund.* **4**, 445–451 (1965).

[17] Hedenqvist M.S., Ritums J.E., Conde-Bran M. and Bergman G., Sorption and sorption of tetrachlorethylene in fluoropolymers : Effects of the chemical structure and crystallinity, *J. Appl. Polymer Sci.* **87**,1474–1483 (2003).

[18] Lisle I.G. and Parlange J-Y., Analytical reduction for a concentration dependent diffusion problem, *ZAMP* **44**, 85-102 (1993).

[19] Libardi P.L., Reichardt K., Jose C., Bazz M. and Nielsen D.R., An approximate method of estimating soil water diffusivity for different soil bulk densities, *Water Resour. Res.* it 18, 177–181 (1982).

[20] Budd Ch.J. and Stockie J.M., Asymptotic behaviour of wetting fronts in porous media with exponential moisture diffusivity, *University of Bath report*, University of Bath, 2012.

[21] Sadighi A. and Ganji D.D., Exact solutions of nonlinear diffusion equations by variational iteration method, *Comp. Math Appl.* **54**, 1112–1121 (2007).

[22] Hristov J., An approximate analytical (integral-balance) solution to a nonlinear heat diffusion equation, *Thermal Science* **19**, 723–733 (2015).

[23] Hristov J., Integral solutions to transient nonlinear heat (mass) diffusion with a power-law diffusivity: a semi-infinite medium with fixed boundary conditions, *Heat Mass Transfer* **52**, 635–655 (2016).

[24] Vazquez J.L., *The porous media equation: Mathematical Theory*. Oxford Mathematical Monographs, Clarendon Press, Oxford, 2007.

[25] Crank J., A theoretical investigation of the influence of molecular relaxation and internal stress on diffusion in polymers, *J.Polym. Sci.* **11**, 151–168 (1953).

[26] Witelski T.P., Horizontal infiltration into wet soil, *Water Resour. Res.* **34**, 1859–1863 (1998).

[27] Layreya aK.B. and Micksoon A., Scaling the exponential soil water diffusivity for soils from Ghana, *J. Hydrology* **79**, 95–106 (1985

[28] Zhou C, General solution of hydraulic diffusivity from sorptivity test, *Cement and Concrete Res.* **58**, 152–160 (2014).

[29] Laurence R.L. and Slattery J.C, Diffusion in Ethylene-Propylene rubber, *J. Polym. Sci.* **5**, 1327–1340 (1967).

[30] Helmroth I.E., Dekker M. and Hankemeier Th., Additive diffusion from LDPE slabs into contacting solvents as a function of solvent absorption, *J. Appl. Polym. Sci.* **90**, 1609–1617 (2003).

[31] Zhou S. and Stern A., The effect of plastifization on the transport of gases in and through glassy polymers, *J.Polym. Sci.:Part B* **27**, 205–222 (1989).

[32] Clothier B.E. and White I., Measurement of sorptivity and soil-water diffusivity in the field, *Soil. Sci. Soc. Am. J.* **45**, 241–245 (1981)

[33] Miller R.D. and Bressler E., A quick method for estimating soil-water diffusivity functions, *Soil. Sci. Soc. Am. J.* **41**, 1020–1022 (1977).

[34] Paralange, J.-Y. and Braddock R.D., An applications of Brutsaerts and optimization techniques to the nonlinear diffusion equation: the Influence of tailing, *Soil Science* **129**, 145–149 (1980).

[35] Tolikas, P.K, Tzimopulos, Ch.D. and Tolikas, D.K., A simple analytical solution for horizontal absorption of water by soils for exponential soil water diffusivity, *Water Resour. Res.* **16**, 821825 (1980).

[36] Tzimopulos Ch., Evangelides Ch. and Arampatzis G., Explicit approximate solution of the horizontal diffusion equation, *Soil Science* **180**, 47–3 (2015).

[37] Liu, P.L-F., A perturbation solution for a nonlinear diffusion equation, *Water Resour. Res*, **12**, 1235–1240 (1976).

[38] Riek R.F., McAvoy T.J. and Chapepelear D.C., Diffusion in polymer-solvent systems. A study of numerical methods of simulations, *J. Polym.Sci., A2-Polymer Physics* **6**, 1863–1886 (1968).

[39] Babu D.K., Infiltration analysis and perturbation methods. 1. Absorption with exponential diffusivity, *Water Resouces Res.* **12**, 80–93 (1976).

[40] Parlange J-Y., Theory of water-movement in soils: 1.One-dimensional absorption, *Soil Science* **111**, 134–137 (1971).

[41] Bruce R.R. and Klute A., The measurement of soil moisture diffusivity, *Soil Science Society of America Proceedings* **20**, 458–462 (1956).

[42] Parlange, J.-Y. and Babu D.K., A comparison of techniques for solving the diffusion equation with an exponential diffusivity, *Water Resour. Res.* **12**, 13171318 (1976).

[43] Duda J.L. and Vrentas J.S., Mathematical analysis of sorption experiments, *AIChE J.* **17**, 464–469 (1971).

[44] Ames W.F., *Nonlinear partial differential equations in engineering*. Academic Press, New York., 1965.

[45] Goodman T.R., Application of Integral Methods to Transient Nonlinear Heat Transfer, *Advances in Heat Transfer*, T. F. Irvine and J. P. Hartnett, eds., 1 (1964), Academic Press, San Diego, CA, pp. 51122.

[46] Hristov J., The heat-balance integral method by a parabolic profile with unspecified exponent: Analysis and Benchmark Exercises, *Thermal Science* **13**, 27–48 (2009).

[47] Sadoun N., Si-Ahmed E.K. and Colinet P., On the refined integral method for one-phase Stefan problem with time-dependent boundary conditions, *Appl. Math. Model.* **30**, 531–544 (2006).

[48] Mitchell S.L. and Myers T.G., Application of standard and refined heat balance integral methods to one-dimensional Stefan problems, it SIAM Review **52**, 57–86 (2010).

[49] Volkov V.N. and Li-Orlov V.K., A Refinement of the Integral Method in Solving the Heat Conduction Equation, *Heat Transfer Sov. Res.* **2**, 41–47 (1970).

[50] Hristov J. The Heat-Balance Integral: 1. How to Calibrate the Parabolic Profile?, *Comptes Rendues Mechanique* **340**, 485–492 (2012).

[51] Myers T.G., Optimizing the exponent in the heat balance and refined integral methods, *Int. Comm. Heat Mass Transfer* **36**, 143–147 (2009).

[52] Lockington D., Parlange J-Y. and Doux P., Sorptivity and the estimation of water penetration into unsaturated concrete, *Mater. Sruct.* **32**, 342–347 (1999).

[53] Hall C., Water sorptivity of mortars and concretes: a review, *Mag.Concrete Res.* **41**, 51–61 (1989

[54] Parlange M.B, Prasad S.N., Parlange J-Y. and Romkens M., Extension of the Heaslet-Alksne technique to arbitrary soil-waterdiffusivities, *Water Resour. Res.* **28**, 2793–2797 (1992).

[55] Tzimopoulos Ch., Evangelides Ch. and Arampatzis G., Explicit approximate analytical solution of the horizontal diffusion equation, *Soil Science* **180**, 47–53 (2015).

[56] Philip J.R. and Knight J.H., On solving the unsaturated flow equation: 3 New quasi-analytical technique, *Soil Science* **117**, 1–13 (1974).

In: A Closer Look at the Diffusion Equation
Editor: Jordan Hristov
ISBN: 978-1-53618-330-6
© 2020 Nova Science Publishers, Inc.

Chapter 4

SOLUTIONS FOR FRACTIONAL REACTION-DIFFUSION EQUATIONS

D. Marin[1,*], *L. M. S. Guilherme*[1,†], *M. K. Lenzi*[2,‡],
E. K. Lenzi[1,3,§] *and P. M. Ndiaye*[4,¶]

[1]Departamento de Física, Universidade Estadual de Ponta Grossa,
Ponta Grossa, Paraná, Brazil
[2]Departamento de Engenharia Química,
Universidade Federal do Paraná, Curitiba, PR, Brazil
[3]National Institute of Science and Technology for Complex Systems,
Centro Brasileiro de Pesquisas Físicas, Rio de Janeiro, RJ, Brazil
[4]Departamento de Engenharia Química, Escola de Química,
Universidade Federal do Rio de Janeiro - Rio de Janeiro, RJ, Brazil

Abstract

We analyze the solutions of different fractional reaction-diffusion equations, which can be related to irreversible or reversible processes. They can be obtained from the continuous-time random walk approach by considering suitable conditions connected to the reaction processes, e.g., of remotion of the walkers, in the case of irreversible processes, or

---

[*]Corresponding Author's Email: daramarin@hotmail.com.
[†]Corresponding Author's Email: lmarcelolj@hotmail.com.
[‡]Corresponding Author's Email: lenzi@ufpr.br.
[§]Corresponding Author's Email: eklenzi@uepg.br.
[¶]Corresponding Author's Email: papa@eq.ufrj.br.

on the probability density function to incorporate different fractional differential operators. For them, we obtain exact solutions in terms of the Green function approach and show that they can be related to a rich variety of behavior related to anomalous diffusion.

Keywords: radiation heat transfer, heat conduction, memory, fractional derivatives

1. INTRODUCTION

Sorption - desorption and reaction processes present in biophysical and biochemical phenomena play an important role in the mechanisms of the living systems. They are essential for the life existence and are also directly influenced by the surfaces, which in general are selective membranes, and the substrate, where the diffusion and the transport occur among the parts of the system [1, 2, 3, 4, 5]. Thus, we have an interplay among these processes and their various mechanisms, which leads us to a challenging problem of how to describe these systems.

Different approaches have been proposed, from the mathematical point of view, to capture the nature of these processes in connection with the experimental results [6, 7, 8, 9, 10, 11] in order to investigate the mechanisms present in these processes. The fractional calculus is one of the approaches used in many situations [12, 13, 14, 15, 16, 17]. It is a simple way of incorporating different effects, which are not present in the standard differential operators, by extending the integer order of the differential or integral equations to a non-integer order. It has been considered in several situations such as diffusion on fractals, drug absorption [18, 19] (see also Ref. [20]), tumor growth and anti-cancer effects [21], bioengineering [22, 23, 24], diffusion and drug deliver [19], kinetics in spiny dendrites [25], charge carriers in organic semiconductors [26], chemotaxis [27], and crowded environments [28]. Situations with different regimes of diffusion, e.g., from sub-diffusion to usual diffusion [29], have also been investigated with this approach by considering fractional diffusion equations of distributed order. It is worth mentioning that in these systems the relaxation processes are non-Debye, i.e., are not exponential, and the diffusion is not standard with the mean square displacement manifesting a nonlinear time dependence, e.g., $\left\langle (x - \langle x \rangle)^2 \right\rangle \sim t^\alpha$, where $\alpha < 1$ and $\alpha > 1$ correspond to subdiffusion and superdiffusion, respectively. These features are typical of non-Markovian processes, which may be related to memory effects, long-range correlations, and long-range interactions. Other extensions, such as the nonlinear diffusion equations [30], have also been considered to investigate reaction-diffusion processes.

Here, we analyze the solutions for different fractional reaction-diffusion equations used to modeling reaction-diffusion processes. Our analysis starts with the equations developed in Ref. [31], which were obtained from the continuous-time random walks approach with non-instantaneous and instantaneous annihilation processes. Following, we extend and analyze one of the formulations developed in Ref. [31] to an arbitrary integrodifferential operator, which may cover different scenarios characterized by singular and non-singular (or mixing of them) fractional operators, leading us to a rich class of behaviors. These developments are performed in Sec. II, by considering irreversible and reversible reaction processes and by using the Green function approach to obtain the solutions. In Sec. III, we present our discussions and conclusions.

2. REACTION - DIFFUSION

Let us consider the reaction-diffusion problem in the context of the fractional diffusion equations by considering different approaches and after performing a comparison among them. We first consider an irreversible process and after a reversible process, where different species may be present.

2.1. Irreversible Reaction

We start our analysis of these approaches by reviewing the cases characterized by a simple reaction process, i.e., an irreversible process connected to a kinetic equation of first order. This scenario can be related to different situations for which the substrate removes or immobilizes the particles which diffuse. The case to be analyzed is governed by the following equation:

$$\frac{\partial}{\partial t}\rho(x,t) = \mathcal{D}_\gamma e^{-\alpha t}\,{}^{RL}_{\ 0}D^{1-\gamma}_t \left\{ e^{\alpha t} \frac{\partial^2}{\partial x^2}\rho(x,t) \right\} - \alpha \rho(x,t) \qquad (1)$$

where α is the reaction rate, $\mathcal{D}_\gamma$ is the diffusion coefficient, and the fractional operator in Eq. (1) is the Riemann-Liouville operator, which is defined as follows:

$$^{RL}_{\ 0}D^{1-\gamma}_t \rho(x,t) = \frac{1}{\Gamma(\gamma)} \frac{\partial}{\partial t} \int_0^t dt' \frac{\rho(x,t')}{(t-t')^{1-\gamma}} \qquad (2)$$

for $0 < \gamma < 1$. Equation (1) was obtained in Ref. [31] in terms of the random walk approach [32] by considering non-instantaneous annihilation process and was analyzed by considering different situations. The solution for this case, in terms of the Green function approach [33], is given by

$$\rho(x,t) = e^{-\alpha t} \int_{-\infty}^{\infty} dx' \varphi(x') \mathcal{G}_f^{(1)}(x-x',t) \tag{3}$$

for the initial condition $\rho(x,0) = \varphi(x)$ (where $\varphi(x)$ is an arbitrary normalized function) and the boundary condition $\rho(\pm\infty,t) = 0$. In Eq. (3), $\mathcal{G}(x,t)$ is the Green function related to the free case, i.e.,

$$\mathcal{G}_f^{(1)}(x,t) = \frac{1}{\sqrt{4\pi D_\gamma t^\gamma}} H_{1,1}^{1,0}\left[\frac{|x|}{\sqrt{D_\gamma t^\gamma}}\bigg|_{(0,1)}^{(\frac{1}{2}+n\gamma,\frac{1}{2})}\right], \tag{4}$$

where $H_{p,q}^{m,n}\left[x\big|_{(b,B)}^{(a,A)}\right]$ is the H Fox function [34]. Equation (4) may be obtained by solving the following fractional equation:

$$\frac{\partial}{\partial t}\mathcal{G}_f^{(1)}(x,t) - {}_0^{RL}D_t^{1-\gamma}\left\{\frac{\partial^2}{\partial x^2}\mathcal{G}_f^{(1)}(x,t)\right\} = \delta(x)\delta(t) \tag{5}$$

with the conditions $\mathcal{G}(\pm\infty,t) = 0$ and $\mathcal{G}(x,t) = 0$ for $t < 0$. In order to solve this equation, we may use the Laplace transform to simplify the calculations. In this sense, after applying the Laplace transform in Eq. (5), we obtain that

$$s\mathcal{G}_f^{(1)}(x,s) - s^{1-\gamma}D_\gamma \frac{\partial^2}{\partial x^2}\mathcal{G}_f^{(1)}(x,s) = \delta(x) \tag{6}$$

whose solution, after some calculations, can be written as

$$\mathcal{G}_f^{(1)}(x,s) = \frac{1}{s\sqrt{D_\gamma/s^\gamma}} e^{\sqrt{\frac{s^\gamma}{D_\gamma}}|x|}. \tag{7}$$

The inverse of Laplace transform of the previous equation yields Eq. (4) by using the result:

$$\mathcal{L}^{-1}\left\{s^{-\alpha}e^{-as^\sigma}\right\} = t^{\alpha-1}H_{1,1}^{1,0}\left[at^{-\sigma}\bigg|_{(0,1)}^{(\alpha,\sigma)}\right], \tag{8}$$

presented in Ref. [34]. By considering the previous equations, it is possible to find the mean square displacement related to this case and, consequently, quantify the time dependence of the spreading of the system. In particular, it is, for simplicity, with $\rho(x,0) = \delta(x)$, given by

$$\begin{aligned}\sigma_x^2(t) &= \langle (x - \langle x \rangle)^2 \rangle \\ &= \frac{2\mathcal{D}_\gamma t^\gamma}{\Gamma(1+\gamma)} e^{-k_1 t},\end{aligned} \qquad (9)$$

which implies that for small times the spreading of the system is anomalous and for a long time decays exponentially. It is also interesting to obtain the time behavior of the quantity of substance (or specie) present in the bulk, i.e., $\mathcal{S}(t) = \int_{-\infty}^{\infty} dx \rho(x,t)$. For this case, this quantity is given by $\mathcal{S}(t) = e^{-k_1 t}$ for the initial condition $\mathcal{S}(0) = 1$, which implies in an exponential relaxation. This feature shows that the systems governed by Eq. (1) have an anomalous spreading of the distribution, while the reaction term acts removing or immobilizing the particles with an exponential rate. The last term is responsible for the exponential decay of $\mathcal{S}(t)$, which can be related to a standard kinetic equation of first order.

Another formulation used in connection with fractional differential operators is based on the following equation:

$$\frac{\partial^\gamma}{\partial t^\gamma} \rho(x,t) = \mathcal{D}_\gamma \frac{\partial^2}{\partial x^2} \rho(x,t) - \alpha \rho(x,t) \qquad (10)$$

where the fractional operator present in Eq. (10) is in Caputo sense and is defined as follows:

$$\frac{\partial^\gamma}{\partial t^\gamma} \rho(x,t) = \frac{1}{\Gamma(\gamma)} \int_0^t \frac{dt'}{(t-t')^{1-\gamma}} \frac{\partial}{\partial t'} \rho(x,t') \qquad (11)$$

for $0 < \gamma < 1$. Note that an equivalent form for Eq. (10) was obtained in Ref. [31] by considering the continuous random walk approach with instantaneous annihilation process. In addition, an extension of Eq. (10) has been worked out in Refs. [35, 30] by considering external forces and a nonlocal term.

The solution for Eq. (10) can be found by using the Green function approach and integral transforms as performed in the last case. Before using the Green function approach, we may apply the Laplace transform in Eq. (10), yielding

$$s\rho(x,s) - \varphi(x) = s^{1-\gamma} \mathcal{D}_\gamma \frac{\partial^2}{\partial x^2} \rho(x,s) - s^{1-\gamma} \alpha \rho(x,s) \qquad (12)$$

with the solution formally given by

$$\rho(x, s) = \int_{-\infty}^{\infty} dx' \varphi(x') \mathcal{G}_{k_r}^{(2)}(x - x', s; \alpha) , \qquad (13)$$

where the Green function for this case, related to Eq. (12), is obtained by solving the equation

$$s\mathcal{G}_{k_r}^{(2)}(x, s; k_r) - s^{1-\gamma}\mathcal{D}_\gamma \frac{\partial^2}{\partial x^2}\mathcal{G}_{k_r}^{(2)}(x, s; k_1) + s^{1-\gamma}\alpha\mathcal{G}_{k_r}^{(2)}(x, s; k_r) = \delta(x) . \qquad (14)$$

subjected to the conditions $\mathcal{G}(\pm\infty, t) = 0$ and $\mathcal{G}(x, t) = 0$ for $t < 0$. The Fourier transform applied in Eq. (14) yields

$$\mathcal{G}_{k_r}^{(2)}(k, s; k_r) = \frac{s^{\gamma-1}}{s^\gamma + \mathcal{D}_\gamma k^2 + \alpha} . \qquad (15)$$

Now, before performing the inverse of Laplace transform of this Green function, we perform an expansion in order to write it as follows:

$$\mathcal{G}_{k_r}^{(2)}(k, s; \alpha) = \sum_{n=0}^{\infty} (-k_1)^n \frac{s^{\gamma-1}}{(s^\gamma + \mathcal{D}_\gamma k^2)^{n+1}} , \qquad (16)$$

which enable us to use the result [34]:

$$\mathcal{L}^{-1}\left\{\frac{s^{\alpha-\beta}}{(s^\alpha \pm a)^{n+1}}\right\} = \frac{1}{n!} t^{\alpha n + \beta - 1} E_{\alpha,\beta}^{(n)}(\mp a t^\alpha) \qquad (17)$$

and, consequently, obtain

$$\mathcal{L}^{-1}\left\{\frac{s^{\gamma-1}}{(s^\gamma + \mathcal{D}_\gamma k^2)^{n+1}}\right\} = \frac{1}{n!} t^{n\gamma} E_\gamma^{(n)}(-\mathcal{D}_\gamma k^2 t^\gamma) , \qquad (18)$$

where $E_{\alpha,\beta}(x)$ is a generalized Mittag-Leffler function (for $\beta = 1$ it recovers the Mittag - Leffler function $E_\gamma(x)$) and $E_{\alpha,\beta}^{(n)}(x)$ is defined as follows

$$E_{\alpha,\beta}^{(n)}(x) = \frac{d^n}{dx^n} E_{\alpha,\beta}(x) = n! \, E_{\gamma,\beta+\gamma n}^{n+1}(x)$$

where $E_{\gamma,\beta}^\delta(x)$ is the three parameter Mittag-Leffler function defined as

$$E_{\gamma,\beta}^\delta(x) = \sum_{k=0}^{\infty} \frac{(\delta)_k}{\Gamma(\gamma k + \beta)} \frac{x^k}{k!} ,$$

with $(\delta)_k = \Gamma(\delta + k)/\Gamma(\delta)$. These results allow us to write the Green function in Fourier space as

$$\mathcal{G}^{(2)}_{k_r}(k, s; \alpha) = \sum_{n=0}^{\infty} \frac{(-\alpha t^\gamma)^n}{\Gamma(1+n)} E^{(n)}_\gamma \left(-k^2 t^\gamma\right), \qquad (19)$$

which in terms of the Fox H function can be rewritten as

$$\mathcal{G}^{(2)}_{k_r}(k, s; \alpha) = \sum_{n=0}^{\infty} \frac{(-\alpha t^\gamma)^n}{\Gamma(1+n)} H^{1,1}_{1,2}\left[\mathcal{D}_\gamma k^2 t^\gamma \Big|_{(0,1)(-n\gamma,\gamma)}^{(0,1)}\right], \qquad (20)$$

where the following formulae was used

$$E^{(n)}_{\alpha,\mu}(x) = H^{1,1}_{1,2}\left[-x \Big|_{(0,1)(1-(\alpha n+\mu),\alpha)}^{(0,1)}\right]. \qquad (21)$$

By applying the inverse Fourier transform in Eq. (20) and by using some properties of the Fox H function [30], we obtain that

$$\mathcal{F}^{-1}\left\{H^{1,1}_{1,2}\left[\mathcal{D}_\gamma k^2 t^\gamma \Big|_{(0,1)(-n\gamma,\gamma)}^{(0,1)}\right]\right\} = \frac{1}{\sqrt{4\mathcal{D}_\gamma t^\gamma}} H^{1,0}_{1,1}\left[\frac{|x|}{\sqrt{\mathcal{D}_\gamma t^\gamma}} \Big|_{(0,1)}^{\left(\frac{1}{2}+n\gamma, \frac{1}{2}\right)}\right]$$

and, consequently,

$$\mathcal{G}^{(2)}_{k_r}(x, t; \alpha) = \frac{1}{\sqrt{4\mathcal{D}_\gamma t^\gamma}} \sum_{n=0}^{\infty} \frac{(-\alpha t^\gamma)^n}{\Gamma(1+n)} H^{1,0}_{1,1}\left[\frac{|x|}{\sqrt{\mathcal{D}_\gamma t^\gamma}} \Big|_{(0,1)}^{\left(\frac{1}{2}+n\gamma, \frac{1}{2}\right)}\right]. \qquad (22)$$

By using the previous equations, it is possible to obtain the mean square displacement related to this case. It is, for simplicity, with $\rho(x, 0) = \delta(x)$, given by

$$\sigma_x^2(t) = 2\mathcal{D}_\gamma t^\gamma E^{(1)}_{\gamma,\gamma}(-\alpha t^\gamma), \qquad (23)$$

which is different from the previous one given by Eq. (9). In this context, it is also interesting to obtain the behavior of $\mathcal{S}(t)$ to analyze the time dependence manifested by the sorption process. After some calculations, it is possible to show that $\mathcal{S}(t) = E_\gamma(-\alpha t^\gamma)$, which is asymptotically characterized by a power-law decay. In fact, the asymptotic behavior of the Mittag-Leffler function is $E_\gamma(x) \sim -1/\left[\Gamma(1+\gamma)x\right]$, which implies in $\mathcal{S}(t) \sim 1/(\alpha t^\gamma)$ for long times.

In Figs. (1) and (2), we illustrate the behavior of these equations and the mean square displacement related to each one.

The formulation based on the following equation:

$$\frac{\partial}{\partial t}\rho(x,t) = \mathcal{D}_\gamma \,_0D_t^{1-\gamma}\left\{\frac{\partial^2}{\partial x^2}\rho(x,t)\right\} - \alpha\rho(x,t) \qquad (24)$$

may lead to a non-physical behavior for the solutions as shown in Ref. [31]. However, it is interesting to note that depending of the choice of reaction term, extensions based on Eq. (24) may lead us to obtain non-negative solutions, see Ref [30]. One of this scenario is, for example,

$$\frac{\partial}{\partial t}\rho(x,t) = \mathcal{D}_\gamma \,_0D_t^{1-\gamma}\left\{\frac{\partial^2}{\partial x^2}\rho(x,t)\right\} - \int_0^t dt'\,\alpha(t-t')\rho(x,t') \qquad (25)$$

with a suitable choice of $\alpha(t)$.

Following our discussion about fractional diffusion equations and reaction terms, it is worth mentioning that Eq. (10), as mentioned before, may equivalently be rewritten as follows (see Ref. [31]):

$$\frac{\partial}{\partial t}\rho(x,t) = \mathcal{D}_\gamma \,_0^{RL}D_t^{1-\gamma}\left\{\frac{\partial^2}{\partial x^2}\rho(x,t) - \alpha\rho(x,t)\right\} \qquad (26)$$

in terms of the Riemann-Liouville fractional operator and it is very different from Eq. (24). The interesting point about Eq. (26) is the possibility of considering a different fractional differential operator, instead of the fractional Riemann-Liouville operator. In fact, Eq. (26) may be extended to a different scenarios by considering the presence of a general integro-differential operator. In this case, Eq. (26) could be written as

$$\frac{\partial}{\partial t}\rho(x,t) = \mathcal{D}_\gamma \mathfrak{F}_t^\gamma\left\{\frac{\partial^2}{\partial x^2}\rho(x,t) - \alpha\rho(x,t)\right\}. \qquad (27)$$

In Eq. (27), $\mathfrak{F}_t^\alpha\{\cdots\}$ represents a fractional operator and it is defined as follows:

$$\mathfrak{F}_t^\gamma\{\rho(x,t)\} = \frac{\partial}{\partial t}\int_0^t \rho(x,t')\mathfrak{K}_\gamma(t-t')dt', \qquad (28)$$

where the choice of the kernel $\mathfrak{K}_\gamma(t)$ defines the integro-differential operator to be considered in Eq. (27). Note that Eq. (27) may be obtained, for example, from the continuous time random walk approach by considering an arbitrary waiting time distribution related to a probability density function connected to the dynamics of the random walk. In this sense, it is worth mentioning that Eq. (28) enables us to consider a wide class of scenarios with singular or non-singular kernels in a unified way, which leads us to different relaxation processes for the solutions of Eq. (27). Equation (26) is recovered with the choice

$$\mathfrak{K}_\gamma(t) = t^{\gamma-1}/\Gamma(\gamma), \quad (29)$$

which can be directly related to the well-known Riemann – Liouville fractional operator [30] for $0 < \gamma < 1$. Another choices are also possible such as the exponential kernels. In this sense, it could be considered the kernel [36, 37, 38]

$$\mathfrak{K}_\gamma(t) = \mathcal{R}(\gamma)\exp(-\gamma' t), \quad (30)$$

where $\gamma' = \gamma/(1-\gamma)$ and $\mathcal{R}(\gamma)$ is a normalization constant. This kernel has been related to a different physical situations, in particular the resetting processes [38]. The solution of Eq. (27), by considering the kernel given by Eq. (30), can be also given in terms of Eq. (13) with the Green function given by:

$$\begin{aligned}\mathcal{G}^{(3)}_{k_r}(x,t;\alpha) &= \frac{1}{\sqrt{4\pi \mathcal{D}_\gamma \mathcal{R}(\gamma)t}} e^{-(\gamma'+\mathcal{R}(\gamma)\alpha)t'} e^{-\frac{x^2}{4\mathcal{D}_\gamma \mathcal{R}(\gamma)}} \\ &+ \int_0^t \frac{dt'}{\sqrt{4\pi \mathcal{D}_\gamma \mathcal{R}(\gamma)t'}} e^{-(\gamma'+\mathcal{R}(\gamma)\alpha)t'} e^{-\frac{x^2}{4\mathcal{D}_\gamma \mathcal{R}(\gamma)t'}}. \end{aligned} \quad (31)$$

The mean square displacement, for simplicity, with $\rho(x,0) = \delta(x)$, related to this case is given by

$$\sigma_x^2(t) = \frac{2\mathcal{D}_\gamma \mathcal{R}(\gamma)}{\gamma' + \alpha \mathcal{R}(\gamma)}\left(1 - e^{-(\gamma'+\alpha\mathcal{R}(\gamma))t}\right), \quad (32)$$

which for long times, i.e., $t \to \infty$, is constant, i.e., $\sigma_x^2(t) \to const.$ The behavior of the quantity of substance (specie or particles) present in the bulk is governed by the following equation:

$$\mathcal{S}(t) = e^{-(\gamma'+\mathcal{R}(\gamma)\alpha)t'} + \frac{\gamma'}{\gamma' + \mathcal{R}(\gamma)\alpha}\left(1 - e^{-(\gamma'+\mathcal{R}(\gamma)\alpha)t'}\right). \quad (33)$$

This result is very different of the previous ones in the sense that an stationary

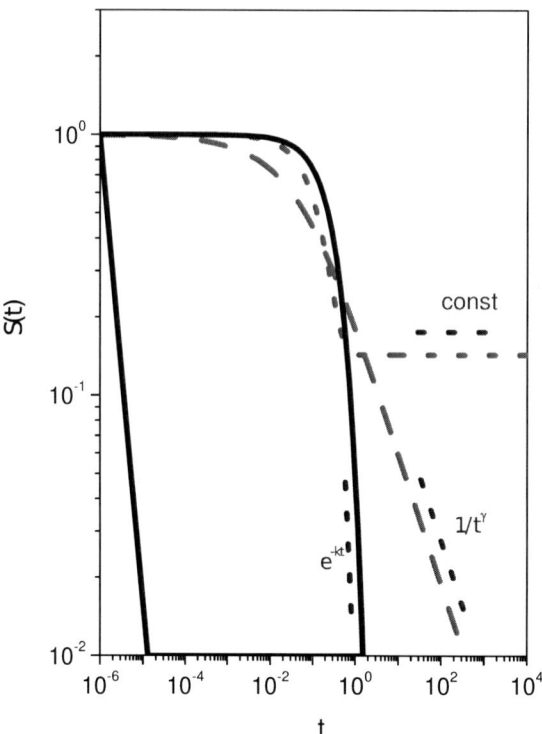

Figure 1. Behavior of $\mathcal{S}(t)$ versus t for different reaction diffusion processes. The green dotted line corresponds to Eq. (1) with an exponential decay. The red dashed line corresponds to the reaction process governed by Eq. (27) for an exponential kernel. The black solid line corresponds to the reaction process obtained from Eq. (39). We consider, for simplicity, $\alpha = 3$, $\mathcal{D}_\gamma = 1$, and $\gamma = 1/2$. Note that for the exponential kernel an stationary, i.e., a time independent behavior, is observed. The blue dotted lines were incorporated in order to show the asymptotic behavior for long times.

solution is present for long times, i.e., $t \to \infty$ implies in $\mathcal{S}(t) \to const$. Figures 1 and 2 illustrate Eqs. (32) and (33) and show the stationary behavior for long times. This point may be applicable for a subtract which may modify the relaxation process of the spreading of the substance (or specie) and presents a saturation. The kernel

$$\mathfrak{K}_\gamma(t) = \mathcal{R}(\gamma) E_\gamma\left(-\gamma' t^{\gamma'}\right), \tag{34}$$

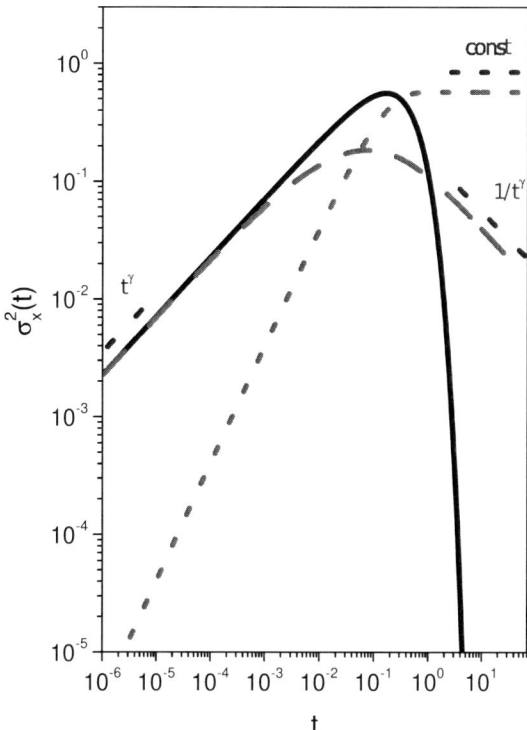

Figure 2. Behavior of σ_x^2 versus t for different kernels. The black line corresponds to Eq. (9) with an exponential decay. The red dashed line corresponds to the reaction process obtained from Eq. (23). The green dotted line corresponds to the reaction process governed by Eq. (32) for an exponential kernel. We consider, for simplicity, $\alpha = 3$, $\mathcal{D}_\gamma = 1$, and $\gamma = 1/2$. Note that for the Eq. (32) an stationary, i.e., a time independent behavior, is observed. The blue dotted lines were incorporated in order to show the asymptotic behavior for small and long times.

has also been used in different applications and can be related to situations characterized by different regimes of diffusion. An interesting point about this kernel is the power-law behavior obtained for $t \to \infty$ since that the asymptotic behavior of the Mittag-Leffler function is characterized by a power-law. The solution for this case can also be found by using the previous approach based on the Green function approach. For this case the Green function is given by

$$\mathcal{G}_{kr}^{(3)} \quad (x,t;\alpha) = \Upsilon(x,t) + \sum_{n=1}^{\infty}(-1)^n \int_{-\infty}^{\infty} dx_n \int_0^t dt_n \Phi(x-x_n, t-t_n) \cdots$$

$$\times \int_{-\infty}^{\infty} dx_2 \int_0^{t_3} dt_2 \Phi(x_3-x_2, t_3-t_2) \int_{-\infty}^{\infty} dx_1 \int_0^{t_2} dt_1 \Phi(x_2-x_1, t_2-t_1) \Upsilon(x_1, t_1) \quad (35)$$

with

$$\Upsilon(x,t) = \frac{1}{\sqrt{4\mathcal{R}(\gamma)\mathcal{D}_\gamma t}} e^{-\frac{x^2}{4\mathcal{R}(\gamma)\mathcal{D}_\gamma t}} + \frac{\gamma' t^\gamma}{\sqrt{4\mathcal{R}(\gamma)\mathcal{D}_\gamma t}} H_{1,1}^{1,0}\left[\frac{|x|}{\sqrt{\mathcal{R}(\gamma)\mathcal{D}_\gamma t}}\bigg|_{(0,1)}^{(\frac{1}{2}+\gamma,\frac{1}{2})}\right] \quad (36)$$

and

$$\Phi(x,t) = \alpha\Upsilon(x,t) + \frac{\gamma' t^{\gamma-1}}{\sqrt{4\mathcal{R}(\gamma)\mathcal{D}_\gamma t}} H_{1,1}^{1,0}\left[\frac{|x|}{\sqrt{\mathcal{R}(\gamma)\mathcal{D}_\gamma t}}\bigg|_{(0,1)}^{(\gamma-\frac{1}{2},\frac{1}{2})}\right]. \quad (37)$$

Others generalization, from the previous equation, can be considered. One of them, for example, is the case

$$\frac{\partial}{\partial t}\rho(x,t) = \mathcal{D}_\gamma \mathfrak{F}_t^\gamma \left\{ \frac{\partial^2}{\partial x^2}\rho(x,t) - \int_0^t dt' \alpha(t-t')\rho(x,t') \right\}, \quad (38)$$

where the reaction term is a time dependent function with non-local characteristics, which for $k_1(t) = \alpha\delta(t)$ recovers Eq. (27).

2.2. Reversible Reaction

Now, we consider the fractional diffusion equations by taking into account reaction terms related to a reversible reaction process, i.e., we incorporate a kinetic processes in the previous fractional diffusion equations which is related to process $1 \rightleftharpoons 2$ for the formation of each substance or specie. This process may be related to the set of kinetic equations:

$$\frac{\partial}{\partial t}\rho_1(x,t) = \alpha_2\rho_2(x,t) - \alpha_1\rho_1(x,t) \quad (39)$$

$$\frac{\partial}{\partial t}\rho_2(x,t) = \alpha_1\rho_1(x,t) - \alpha_2\rho_2(x,t), \quad (40)$$

where $\rho_1(x,t)$ and $\rho_2(x,t)$ represent two different species present in the system, i.e., specie 1 and specie 2, and α_1 and α_2 are the reaction rates related to the formation of these species.

Following the developments performed in previous section, we first consider the approach developed in Ref. [39], which is based on the random walk approach for by considering a multi-species scenario. The fractional reaction diffusion obtained in this context (see details in Ref. [39]) is given by

$$\frac{\partial}{\partial t} P(x,t) = \mathcal{D}_\gamma e^{Rt} \, {}_0 D_t^{1-\gamma} \left\{ e^{-Rt} \frac{\partial^2}{\partial x^2} P(x,t) \right\} + R P(x,t) \quad (41)$$

where

$$P(x,t) = \begin{pmatrix} \rho_1(x,t) \\ \rho_2(x,t) \end{pmatrix} \quad \text{and} \quad R = \begin{pmatrix} -\alpha_1 & \alpha_2 \\ \alpha_1 & -\alpha_2 \end{pmatrix}. \quad (42)$$

Note that the equation for a single specie in this context is extended to a general context by considering a matrix formulation. By performing some calculations, it is possible to show that the previous equation (which represents a set of equations) can be written, in terms of the components $\rho_1(x,t)$ and $\rho_2(x,t)$, as follows:

$$\begin{aligned}
\frac{\partial}{\partial t} \rho_1(x,t) &= \tilde{k}_2 \mathcal{D}_\gamma \, {}_0 D_t^{1-\gamma} \left\{ \frac{\partial^2}{\partial x^2} (\rho_1(x,t) + \rho_2(x,t)) \right\} - \alpha_1 \rho_1(x,t) + \alpha_2 \rho_2(x,t) \\
&+ e^{-k_t t} \mathcal{D}_\gamma \, {}_0 D_t^{1-\gamma} \left\{ e^{k_t t} \frac{\partial^2}{\partial x^2} (\tilde{k}_1 \rho_1(x,t) - \tilde{k}_2 \rho_2(x,t)) \right\}
\end{aligned} \quad (43)$$

and

$$\begin{aligned}
\frac{\partial}{\partial t} \rho_2(x,t) &= \tilde{k}_1 \mathcal{D}_\gamma \, {}_0 D_t^{1-\gamma} \left\{ \frac{\partial^2}{\partial x^2} (\rho_1(x,t) + \rho_2(x,t)) \right\} + \alpha_1 \rho_1(x,t) - \alpha_2 \rho_2(x,t) \\
&- e^{-k_t t} \mathcal{D}_\gamma \, {}_0 D_t^{1-\gamma} \left\{ e^{k_t t} \frac{\partial^2}{\partial x^2} (\tilde{k}_1 \rho_1(x,t) - \tilde{k}_2 \rho_2(x,t)) \right\}.
\end{aligned} \quad (44)$$

where $k_t = \alpha_1 + \alpha_2$, $\tilde{k}_1 = \alpha_1/k_t$, and $\tilde{k}_2 = \alpha_2/k_t$ (see Ref. [39] for details the calculations where these equations were obtained). For the initial conditions $\rho_1(x,0) = \varphi_1(x)$ and $\rho_2(x,0) = \varphi_2(x)$, it is possible to show that the solutions for $\rho_1(x,t)$ and $\rho_2(x,t)$, governed by the previous set of equations, are given by

$$\begin{aligned}
\rho_1(x,t) &= \left(\tilde{k}_2 + \tilde{k}_1 e^{-k_t t} \right) \int_{-\infty}^{\infty} dx \varphi_1(x') \mathcal{G}_f^{(1)}(x - x', t) \\
&+ \tilde{k}_2 \left(1 - e^{-k_t t} \right) \int_{-\infty}^{\infty} dx \varphi_2(x') \mathcal{G}_f^{(1)}(x - x', t)
\end{aligned} \quad (45)$$

and

$$\rho_2(x,t) = \tilde{k}_1\left(1-e^{-k_t t}\right)\int_{-\infty}^{\infty} dx\varphi_1(x')\mathcal{G}_f^{(1)}(x-x',t)$$
$$+ \left(\tilde{k}_1+\tilde{k}_2 e^{-k_t t}\right)\int_{-\infty}^{\infty} dx\varphi_2(x')\mathcal{G}_f^{(1)}(x-x',t) \quad (46)$$

where $\mathcal{G}^{(1)}(x,t)$ is the Green function of the free particle case defined in the previous section. The mean square displacement related to the previous set of equations, for simplicity, with $\rho_1(x,0)=\delta(x)$ and $\rho_2(x,0)=\delta(x)$, is given by

$$\sigma_{x,1}^2(t) = \frac{2\mathcal{D}_\gamma t^\gamma}{\Gamma(1+\gamma)}\left(\tilde{k}_2+\tilde{k}_1 e^{-k_t t}\right) + \frac{2\mathcal{D}_\gamma t^\gamma}{\Gamma(1+\gamma)}\tilde{k}_2\left(1-e^{-k_t t}\right) \quad (47)$$

and

$$\sigma_{x,2}^2(t) = \frac{2\mathcal{D}_\gamma t^\gamma}{\Gamma(1+\gamma)}\tilde{k}_1\left(1-e^{-k_t t}\right) + \frac{2\mathcal{D}_\gamma t^\gamma}{\Gamma(1+\gamma)}\left(\tilde{k}_1+\tilde{k}_2 e^{-k_t t}\right). \quad (48)$$

Similar to the previous case, we extend the formulation based on Eq. (10) by considering a reversible process for two different species, as performed above. In this case, we have, for the species $\rho_1(x,t)$ and $\rho_2(x,t)$, the following set of equations:

$$\frac{\partial^\gamma}{\partial t^\gamma}\rho_1(x,t) = \mathcal{D}_\gamma\frac{\partial^2}{\partial x^2}\rho_1(x,t) - \alpha_1\rho_1(x,t) + \alpha_2\rho_2(x,t) \quad (49)$$
$$\frac{\partial^\gamma}{\partial t^\gamma}\rho_2(x,t) = \mathcal{D}_\gamma\frac{\partial^2}{\partial x^2}\rho_2(x,t) + \alpha_1\rho_1(x,t) - \alpha_2\rho_2(x,t) \quad (50)$$

In order to solve the previous set of fractional diffusion equations, we may perform the following change

$$\rho_+(x,t) = \rho_1(x,t) + \rho_2(x,t) \quad \text{and} \quad (51)$$
$$\rho_-(x,t) = \alpha_1\rho_1(x,t) - \alpha_2\rho_2(x,t) \quad (52)$$

in order to simplify Eqs. (49) and (50) and obtain independent fractional diffusion equation to be solved independently each other. In fact, by using $\rho_+(x,t)$ and $\rho_-(x,t)$ and performing some calculations, it is possible to show that Eqs. (49) and (50) can be written in two decoupled equations, which are:

$$\frac{\partial^\gamma}{\partial t^\gamma}\rho_+(x,t) = \mathcal{D}_\gamma\frac{\partial^2}{\partial x^2}\rho_+(x,t), \quad (53)$$

subjected to the initial condition $\rho_+(x,0) = \varphi_1(x) + \varphi_2(x)$ and the boundary condition $\rho_+(\pm\infty, t) = 0$, and

$$\frac{\partial^\gamma}{\partial t^\gamma}\rho_-(x,t) = \mathcal{D}_\gamma \frac{\partial^2}{\partial x^2}\rho_-(x,t) - (\alpha_1 + \alpha_2)\rho_-(x,t) \tag{54}$$

subjected to the initial condition $\rho_-(x,0) = \alpha_1\varphi_1(x) - \alpha_2\varphi_2(x)$ and the boundary condition $\rho_-(\pm\infty, t) = 0$. The solution for Eq. (53) and (54) can be found by using the standard procedures of calculations by employing the Laplace and Fourier transforms as performed for the previous cases. In this sense, by using the Laplace transform in Eq. (53), we obtain

$$s^\gamma \rho_+(k,s) - s^{\gamma-1}\rho_+(k,0) = -\mathcal{D}_\gamma k^2 \rho_+(k,s) \tag{55}$$

which can be written as

$$\rho_+(k,s) = \frac{s^{\gamma-1}}{s^\gamma + \mathcal{D}_\gamma k^2}\rho_+(k,0). \tag{56}$$

By applying the inverse of Laplace transform, we have that

$$\rho_+(k,t) = E_\gamma(-k^2 \mathcal{D}_\gamma t^\gamma)\rho_+(k,0) \tag{57}$$

where $E_\gamma(x)$ is the Mittag-Leffler function. The inverse of Fourier transform of the previous equation yields

$$\rho_+(x,t) = \int_{-\infty}^{+\infty} dx'\, \rho_+(x',0)\mathcal{G}_f^{(1)}(x-x',t) \tag{58}$$

with $\mathcal{G}_f^{(1)}(x,t)$ defined by Eq. (4). For Eq. (54), we will apply the same procedure to find the solution of Eq. (53). Thus, we start by applying the Laplace and Fourier transforms, which lead us to obtain

$$s^\gamma \rho_-(k,s) - s^{\gamma-1}\rho_-(k,0) = -k^2 \mathcal{D}_\gamma \rho_-(k,s) - (\alpha_1 + \alpha_2)\rho_-(k,s)$$

$$\rho_-(k,s) = \frac{s^{\gamma-1}}{s^\gamma + \mathcal{D}_\gamma k^2 + k_t}\rho_-(k,0) \tag{59}$$

Before performing the inverse of Laplace transform, we write the previous equations as follows:

$$\rho_-(k,t) = \sum_{n=0}^\infty (-k_t)^n \frac{s^{\gamma-1}}{(s^\gamma + \mathcal{D}_\gamma k^2)^{n+1}}\rho_-(k,0); \tag{60}$$

which yields

$$\rho_-(k,t) = \sum_{n=0}^{\infty} \frac{(-k_t t^\gamma)^n}{\Gamma(1+n)} E_\gamma^{(n)}\left(-\mathcal{D}_\gamma k^2 t^\gamma\right) \rho_-(k,0) \qquad (61)$$

after applying the inverse Laplace transform. In term of the Fox H function, it can be written as

$$\rho_-(k,t) = \sum_{n=0}^{\infty} \frac{(-k_t t^\gamma)^n}{\Gamma(1+n)} H_{1,2}^{1,1}\left[k^2 \mathcal{D}_\gamma t^\gamma \bigg|_{(0,1)(-n\gamma,\gamma)}^{(0,1)}\right] \rho_-(k,0). \qquad (62)$$

By performing the inverse Fourier transform in previous equation and using some properties of the Fox H function, we obtain that

$$\rho_-(x,t) = \sum_{n=0}^{\infty} \frac{(-k_t t^\gamma)^n}{\sqrt{4\mathcal{D}_\gamma t^\gamma}\Gamma(1+n)} \int_{-\infty}^{+\infty} dx' \rho_-(x',0) H_{1,1}^{1,0}\left[\frac{|x-x'|}{\sqrt{\mathcal{D}_\gamma t^\gamma}} \bigg|_{(0,1)}^{(\frac{1}{2}+n\gamma,\frac{1}{2})}\right] \qquad (63)$$

which can be written as

$$\rho_-(x,t) = \int_{-\infty}^{\infty} dx' \rho_-(x',0) \mathcal{G}_{k_r}^{(2)}(x-x',t;k_t). \qquad (64)$$

where $\mathcal{G}_{k_r}^{(2)}(x,t;k_t)$ is defined by Eq. (22) with $\alpha \to k_t$. By using these results, it is possible to write $\rho_1(x,t)$ and $\rho_2(x,t)$ as follows:

$$\rho_1(x,t) = \tilde{k}_2 \int_{-\infty}^{\infty} dx \left(\varphi_1(x') + \varphi_2(x')\right) \mathcal{G}_f^{(1)}(x-x',t)$$
$$+ \int_{-\infty}^{\infty} dx \left(\tilde{k}_1 \varphi_1(x') - \tilde{k}_2 \varphi_2(x')\right) \mathcal{G}_{k_r}^{(2)}(x-x',t;k_t) \qquad (65)$$

and

$$\rho_2(x,t) = \tilde{k}_1 \int_{-\infty}^{\infty} dx \left(\varphi_1(x') + \varphi_2(x')\right) \mathcal{G}_f^{(1)}(x-x',t)$$
$$- \int_{-\infty}^{\infty} dx \left(\tilde{k}_1 \varphi_1(x') - \tilde{k}_2 \varphi_2(x')\right) \mathcal{G}_{k_r}^{(2)}(x-x',t;k_t) \qquad (66)$$

From these results, it is possible to obtain the mean square displacement related to each specie by using these distributions. They are, for simplicity, with $\rho_1(x,0) = \delta(x)$ and $\rho_2(x,0) = \delta(x)$, given by

$$\sigma_{x,1}^2(t) = \frac{2\tilde{k}_2 \mathcal{D}_\gamma t^\gamma}{\Gamma(1+\gamma)} + \left(\tilde{k}_1 - \tilde{k}_2\right) \mathcal{D}_\gamma t^\gamma E_{\gamma,\gamma}^{(1)}(-k_t t^\gamma) \qquad (67)$$

and

$$\sigma_{x,2}^2(t) = \frac{2\tilde{k}_1 \mathcal{D}_\gamma t^\gamma}{\Gamma(1+\gamma)} - \left(\tilde{k}_1 - \tilde{k}_2\right) \mathcal{D}_\gamma t^\gamma E_{\gamma,\gamma}^{(1)}(-k_t t^\gamma) \tag{68}$$

(see Fig 3).

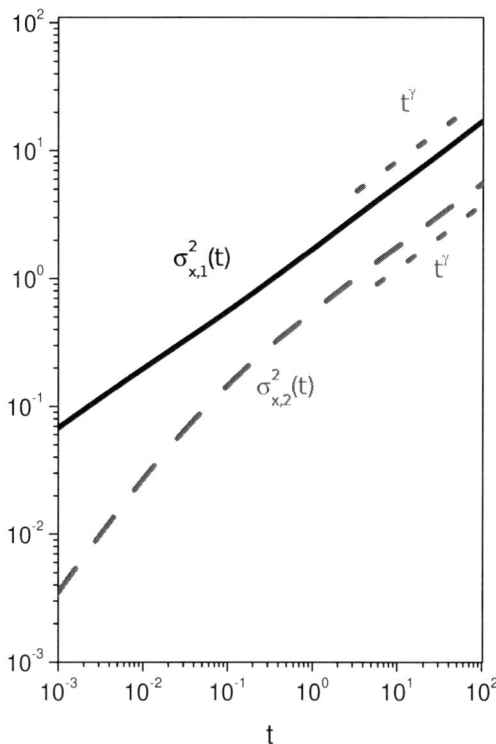

Figure 3. Behavior of σ_x^2 versus t obtained from Eqs. (67) and (68). We consider, for simplicity, $\alpha_1 = 3$, $\alpha_2 = 1$, and $\mathcal{D}_\gamma = 1$. Note that the asymptotic behavior obtained from these equations are the same and can be related to a subdiffusive case with $\gamma = 1/2$. The green dotted lines were incorporated in order to show the asymptotic behavior for long times.

Let us consider another possibility of extending the reaction - diffusion problem by incorporating a reversible reaction process in the context of the fractional diffusion equations given by Eq. (27). This scenario may be characterized by the fractional diffusion equations

$$\frac{\partial}{\partial t}\rho_1(x,t) = \mathfrak{F}_t^\gamma\left\{\mathcal{D}_\gamma\frac{\partial^2}{\partial x^2}\rho_1(x,t) - \alpha_1\rho_1(x,t) + \alpha_2\rho_2(x,t)\right\} \quad (69)$$

and

$$\frac{\partial}{\partial t}\rho_2(x,t) = \mathfrak{F}_t^\gamma\left\{\mathcal{D}_\gamma\frac{\partial^2}{\partial x^2}\rho_2(x,t) + \alpha_1\rho_1(x,t) - \alpha_2\rho_2(x,t)\right\}. \quad (70)$$

By employing the same procedure for the previous case, it is possible simplify the previous set of equations in order to obtain

$$\frac{\partial}{\partial t}\rho_+(x,t) = \mathfrak{F}_t^\gamma\left\{\mathcal{D}_\gamma\frac{\partial^2}{\partial x^2}\rho_+(x,t)\right\} \quad (71)$$

and

$$\frac{\partial}{\partial t}\rho_-(x,t) = \mathfrak{F}_t^\gamma\left\{\mathcal{D}_\gamma\frac{\partial^2}{\partial x^2}\rho_-(x,t) - (\alpha_1 + \alpha_2)\rho_-(x,t)\right\}. \quad (72)$$

In order to obtain the solution for these equations, we also employ the method of Green function and integral transforms. In this sense, the solution for Eq. (71) may be written as

$$\rho_+(x,s) = \int_{-\infty}^{\infty} dx'\rho_+(x',0)\mathcal{G}_f^{(4)}(x-x',t) \quad (73)$$

with the Green function obtained from the equation

$$\frac{\partial}{\partial t}\mathcal{G}_f^{(4)}(x,t) - \mathfrak{F}_t^\gamma\left\{\mathcal{D}_\gamma\frac{\partial^2}{\partial x^2}\mathcal{G}_f^{(4)}(x,t)\right\} = \delta(x)\delta(t) \quad (74)$$

and subjected to the conditions $\mathcal{G}_f^{(4)}(\pm\infty, t) = 0$ and $\mathcal{G}_f^{(4)}(x,t) = 0$ for $t < 0$. In particular, the solution for Eq. (74) is given by

$$\mathcal{G}_f^{(4)}(x,t) = \frac{1}{\sqrt{4\pi\mathcal{D}_\gamma\mathcal{R}(\gamma)t}}e^{-\gamma't'}e^{-\frac{x^2}{4\mathcal{D}_\gamma\mathcal{R}(\gamma)t}} + \int_0^t \frac{dt'}{\sqrt{4\pi\mathcal{D}_\gamma\mathcal{R}(\gamma)t'}}e^{-\gamma't'}e^{-\frac{x^2}{4\mathcal{D}_\gamma\mathcal{R}(\gamma)t'}}. \quad (75)$$

For Eq. (72), we have that the solution is

$$\rho_-(x,s) = \int_{-\infty}^{\infty} dx'\rho_-(x',0)\mathcal{G}_{kr}^{(3)}(x-x',t;k_t). \quad (76)$$

The solutions for this case present an stationary behavior, i.e., for $t \to \infty$ they become time independent, in contrast to the previous cases. From these results, it is possible to obtain the mean square displacement related to these distributions for each specie. It is, for simplicity, with $\rho_1(x,0) = \delta(x)$ and $\rho_2(x,0) = \delta(x)$, given by

$$\sigma_{x,1}^2(t) = \frac{2\alpha_2}{\alpha_1 + \alpha_2} \mathcal{D}_\gamma \mathcal{R}(\gamma) \left(1 - e^{-\gamma' t}\right)$$
$$+ \frac{2\mathcal{D}_\gamma \mathcal{R}(\gamma)}{\gamma' + (\alpha_1 + \alpha_2)\mathcal{R}(\gamma)} \left(1 - e^{-(\gamma' + (\alpha_1 + \alpha_2)\mathcal{R}(\gamma))t}\right) \quad (77)$$

$$\sigma_{x,2}^2(t) = \frac{2\alpha_1}{\alpha_1 + \alpha_2} \mathcal{D}_\gamma \mathcal{R}(\gamma) \left(1 - e^{-\gamma' t}\right)$$
$$- \frac{2\mathcal{D}_\gamma \mathcal{R}(\gamma)}{\gamma' + (\alpha_1 + \alpha_2)\mathcal{R}(\gamma)} \left(1 - e^{-(\gamma' + (\alpha_1 + \alpha_2)\mathcal{R}(\gamma))t}\right). \quad (78)$$

Note that $\sigma_{x,1}^2(t)$ and $\sigma_{x,2}^2(t)$ present an stationary behavior for $t \to \infty$, i.e., $\sigma_{x,1}^2(t) \to const$ and $\sigma_{x,2}^2(t) \to const$ (see, Fig. 4).

Now, we consider Eqs. (71) and (72) with the kernel given Eq. (34), which is non-singular at the origin and asymptotically is governed by a power-law. For this case, the solution is also given by Eq. (73) and (76), with the Green functions given by

$$\mathcal{G}_f^{(4)}(x,x',t) = \frac{e^{-\frac{x^2}{4\pi \mathcal{D}_\gamma t}}}{\sqrt{4\pi \mathcal{D}_\gamma \mathcal{R}(\gamma)t}} + \frac{1}{|x|} \sum_{n=1}^{\infty} \frac{(-\gamma' t^\gamma)^n}{\Gamma(1+n)} H_{2,2}^{2,0}\left[\frac{|x|}{\mathcal{D}_\gamma \mathcal{R}(\gamma)t}\bigg|_{(1,2),(1+n,1)}^{(1+\alpha\gamma,1),(1,1)}\right] \quad (79)$$

and the $\mathcal{G}_f^{(4)}(x,t)$ given by Eq. (35).

3. DISCUSSION AND CONCLUSION

We have analyzed the solutions of different fractional reaction-diffusion equations by considering irreversible and reversible reaction processes. Our analysis started with Eq. (1), which was obtained in Ref. [31] by using the continuous-time random walk and also compared with other models. Here, we have compared it with Eq. (27), which corresponds to consider different fractional differential operators. In particular, we have considered the kernels given by Eqs. (30) and (34), which are non-singular and asymptotically governed by exponential and power-laws. For these fractional diffusion equations, we have obtained the survival probability and the mean square displacement to perform the comparison among them. For the exponential kernel, we have observed a stationary

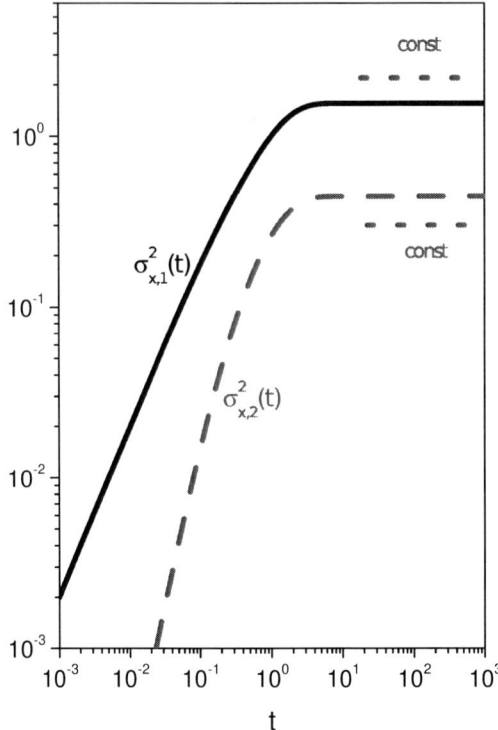

Figure 4. Behavior of σ_x^2 versus t obtained from Eqs. (77) and (78). We consider, for simplicity, $\alpha_1 = 3$, $\alpha_2 = 1$, and $\mathcal{D}_\gamma = 1$. Note that the asymptotic behavior obtained from these equations are the same and can be related to a subdiffusive with $\gamma = 1/2$. The green dotted lines were incorporated in order to show the asymptotic behavior for long times.

solution, i.e., a time-independent solution, in contrast to the other kernels or equations with the Riemann-Liouville or Caputo fractional time derivative. This means that the reaction process, in the presence of this diffusion process, is not able to remove, completely, the species (particles or substance) of the system. We have an interplay between the resetting associated with the exponential kernel and the reaction processes, which reaches a stationary state. Following, we have considered a reversible reaction process by considering two different species and performed a comparison among the different fractional reaction-diffusion equations. This scenario is different from the previous one and leads us to verify that the system asymptotically is governed by the diffusion process. The reaction terms play an important role in intermediate times. Finally, we hope that the results obtained here may be useful for the discussion on the reaction-diffusion processes in different scenarios, where the anomalous diffusion is present.

ACKNOWLEDGMENT

We thank the CNPq and CAPES (Brazilian agencies) for partial financial support.

REFERENCES

[1] Weiss M., Crowding, Diffusion, and Biochemical Reactions, In: New Models of the Cell Nucleus: Crowding, Entropic Forces, Phase Separation, and Fractals. Eds. Ronald Hancock and Kwang W. Jeon. (Academic Press, 2014, vol. 307, pp. 383 - 417).

[2] Burrage K., Hancock J., Leier A., Nicolau Jr D. V., Modelling and simulation techniques for membrane biology, *Briefings in bioinformatics* **8**, 234 - 244 (2007).

[3] Latour L., Svoboda K., Mitra P. P., Sotak C. H., Time-dependent diffusion of water in a biological model system, *P. Natl. Acad. Sci. USA* **91**, 1229 - 1233 (1994).

[4] Fedotov S., Korabel N., Waigh T. A., Han D., Allan V. J., Memory effects and Lévy walk dynamics in intracellular transport of cargoes, *Phys. Rev.E* **98**, 042136 (2018).

[5] Liu X., Sun H.-G., Lazrević M. P., Fu Z., A variable-order fractal derivative model for anomalous diffusion, *Therm. Sci* **21**, 51–59, (2017).

[6] Saxton M. J., Jacobson K., Single-particle tracking: applications to membrane dynamics, *Annu. Rev. Bioph. Biom.* **26**, 373–399, (1997).

[7] Briane V., Vimond M., Kervrann C., An overview of diffusion models for intracellular dynamics analysis, *Briefings in Bioinformatics*, bbz052 (2019).

[8] Qian H., Sheetz M. P., Elson E. L., Single particle tracking. Analysis of diffusion and flow in two-dimensional systems, *Biophys. J.* **60**, 910-921 (1991).

[9] Gillespie D. T., Exact stochastic simulation of coupled chemical reactions, *J. Phys. Chem.* **81**, 2340-2361 (1977).

[10] Havlin S., Ben-Avraham D., Diffusion in disordered media, *Adv. Phys.* **36**, 695–798, (1987).

[11] Bouchaud J.-P., Georges A., Anomalous diffusion in disordered media: statistical mechanisms, models and physical applications, *Phys. Rep.* **195**, 127–293, (1990).

[12] Bressloff P. C., Newby J. M., Stochastic models of intracellular transport, Rev. Mod. Phys. **85**, 135 (2013).

[13] Yang X.-J., Machado J. A., A new fractional operator of variable order: application in the description of anomalous diffusion, *Physica A* **481**, 276-283 (2017).

[14] Sokolov I. M., Klafter J., From diffusion to anomalous diffusion: a century after Einsteins Brownian motion, *Chaos* **15**, 026103 (2005).

[15] Samko S. G., Kilbas A. A., Marichev O. I., Fractional integrals and derivatives : theory and applications, (Gordon and Breach Science Publishers, Yverdon Yverdon-les-Bains, Switzerland, 1993).

[16] Podlubny I., Fractional differential equations: an introduction to fractional derivatives, fractional differential equations, to methods of their solution and some of their applications, (Academic Press, California, 1998).

[17] Metzler R., Klafter J., The restaurant at the end of the random walk: recent developments in the description of anomalous transport by fractional dynamics, *J. Phys. A* **37**, R161 (2004).

[18] Dokoumetzidis A., Macheras P., Fractional kinetics in drug absorption and disposition processes, *J. Pharmacokinet. Pharmacodyn.*, **36**, 165-178, (2009).

[19] Lenzi E. K., Novatski A., Farago P. V., Almeida M. A., Zawadzki S. F., Menechini Neto R., Diffusion Processes and Drug Release: Capsaicinoids - Loaded Poly (ε-caprolactone) Microparticles, *PLOS ONE*, **11**, (2016).

[20] Sopasakis P., Sarimveis H., Macheras P., Dokoumetzidis A., Fractional calculus in pharmacokinetics, *J. Pharmacokinet. Pharmacodyn.* **45**, 107125, (2018).

[21] Atc Ferhan M., Atc Mustafa, Nguyen Ngoc, Zhoroev Tilekbek, Koch Gilbert, A study on discrete and discrete fractional pharmacokinetics-pharmacodynamics models for tumor growth and anti-cancer effects, *Comp. Math. Bioph.* **7**, 10 - 24 (2019).

[22] Magin Richard L., Fractional Calculus in Bioengineering, Part 1, *Critical Reviews & Trade; in Biomedical Engineering*, **32**, (2004).

[23] Magin Richard L., Fractional Calculus in Bioengineering, Part 2, *Critical Reviews & Trade; in Biomedical Engineering*, **32**, (2004).

[24] Magin Richard L., Fractional Calculus in Bioengineering, Part3, *Critical Reviews & Trade; in Biomedical Engineering* **32**, (2004).

[25] Vitali Silvia, Mainardi Francesco and Castellani Gastone, Emergence of Fractional Kinetics in Spiny Dendrites., *Fractal Fract.* **2**, 6 (2018).

[26] Choo K. Y., Muniandy S. V., Woon K. L., Gan M. T., Ong D. S., Modeling anomalous charge carrier transport in disordered organic semiconductors using the fractional drift-diffusion equation, *Organic Electronics* **41**, 157–165 (2017).

[27] Langlands T. A. M., Henry B. I., Fractional chemotaxis diffusion equations, *Phys. Rev. E* **81**, 051102 (2010).

[28] Sokolov I. M., Models of anomalous diffusion in crowded environments, *Soft Matter* **8**, 9043-9052 (2012).

[29] Sun H. G., Li Z., Zhang Y., Chen W., Fractional and fractal derivative models for transient anomalous diffusion: Model comparison, *Chaos, Solitons & Fractals* **102**, 346-353 (2017).

[30] Evangelista L. R. and Lenzi E. K., Fractional Diffusion Equations and Anomalous Diffusion, (Cambridge University Press, 2018).

[31] Henry B. I., Langlands T. A. M. and Wearne S. L., Anomalous diffusion with linear reaction dynamics: From continuous time random walks to fractional reaction-diffusion equations, *Phys. Rev. E*, **74**, 031116, (2006).

[32] Klafter J., Sokolov I. M., First steps in random walks: from tools to applications, (Oxford University Press, 2011).

[33] Wyld H. W., Mathematical Physics (Advanced Book Program, Perseus Books, 1976).

[34] Mathai A. M., Saxena R. K., and Haubold H. J., The H-function: Theory and Applications, (Springer, New York, 2010).

[35] Lenzi E. K., Menechini Neto R., Tateishi A. A., Lenzi M. K., Ribeiro H. V., Fractional diffusion equations coupled by reaction terms, *Physica A* **458**, 9 - 16 (2016).

[36] Caputo M., Fabrizi M., A new definition of fractional derivative without singular kernel, *Progr. Fract. Differ. Appl.* **1**, 73-85 (2015).

[37] Hristov J., Derivation of the Fractional Dodson Equation and Beyond: Transient Diffusion with a Non-Singular Memory and Exponentially Fading-Out Diffusivity, *Progr. Fract. Differ. Appl.* **3**, 255-270, (2017).

[38] Tateishi A. A., Ribeiro H. V., Lenzi E. K., The Role of Fractional Time-Derivative Operators on Anomalous Diffusion, *Frontiers in Physics* **5**, 52 (2017).

[39] Langlands T. A. M., Henry B. I., Wearne S. L., Anomalous subdiffusion with multispecies linear reaction dynamics, *Phys. Rev. E* **77**, 021111 (2008).

In: A Closer Look at the Diffusion Equation
Editor: Jordan Hristov
ISBN: 978-1-53618-330-6
© 2020 Nova Science Publishers, Inc.

Chapter 5

SEMI-ANALYTICAL SOLUTION OF HRISTOV DIFFUSION EQUATION WITH SOURCE

Derya Avcı* and Beyza Billur İskender Eroğlu
Department of Mathematics, Faculty of Science and Letters,
Balikesir University, Balikesir, Turkey

Abstract

The Hristov diffusion models with ABR derivative are based on the relation between AB integral definition and the fading memory concept occurring in the Boltzmann superposition principle. Hence, they have a physically interpretable background, unlike the directly fractionalized models. Motivated on this structure of the Hristov diffusion models, semi-analytical solutions of the first type Hristov diffusion equation with ABR derivative under the effect of space and time dependent external sources are investigated in this chapter. To reduce the problem into the ordinary differential equations, Fourier Method is applied and the resulting time dependent ordinary fractional differential equation is approximated by the Diethelm's predictor-corrector algorithm. Solution procedure and effects of different source functions are illustrated under the variations of the problem parameters.

Keywords: Hristov diffusion, fading memory, ABR derivative, Fourier method, Diethelm's predictor-corrector algorithm

---

*Corresponding Author's Email: dkaradeniz@balikesir.edu.tr.

1. INTRODUCTION

Diffusion phenomena is a multidisciplinary topic in scientific fields including engineering, physics, bio-mathematics, finance and so on. It is a transport of many quantities such as heat, mass, atoms, molecules and ions from the higher concentration to lower concentration [1]. Diffusion starts to move by a gradient in concentration. Hence, the diffusion equation is derived from Fick's first law and called as normal (Fickian) diffusion because of the mean squared displacement of the particles. This characteristic behavior of diffusion is dissipation and so it can be seen in many dissipative systems in the nature. However, experimental results in different applications show that the classical diffusion equation is not the accurate model. Because the Fick's law implies the simultaneous occurrence of cause and effect. For instance in a heat diffusion, when one end of the cable is heated, it is no matter how far from this point, the effect of heating is immediately felt at any point of the cable. This property is unphysical and is not possible for the non-homogeneous or porous materials.

The weakness of the classical diffusion phenomena has led scientists to introduce generalized diffusion models. These models represent the anomalous (non-Fickian) diffusion processes. Among the leading studies on anomalous diffusion, Cattaneo's diffusion model including the relaxation time ξ, a natural modification of Fick's law by revealing the finite propagation speed, is as follows [2]

$$j(x,t) + \xi \frac{\partial j(x,t)}{dt} = -d \frac{\partial c(x,t)}{dx}. \tag{1}$$

Similarly, Cattaneo adopted this idea to the heat diffusion in rigid conductors and generalized the Fourier's law by proposing the relation between heat flux and the history of temperature as

$$q(x,t) = -\int_{-\infty}^{t} R(x,t) \nabla T(x, t-\tau) d\tau, \tag{2}$$

where $R(x,t)$ denotes the damping function.

Besides aiming to generalize the Fick's law, two main reasons for the anomalous diffusion phenomena to be attracted are space-time non-locality and memory effects [3, 4]. At this point, fractional derivatives, which can naturally imply the memory and hereditary effects in many physical processes due to their definitions [5], are of crucial importance. Modeling of anomalous diffusion processes was made until recently with the conventional Riemann-Liouville (RL) and Caputo fractional derivatives [6, 7, 8]. As is known, one of the non-local constitutive relation between flux and concentration leading to fractional diffusion equations is as follows

$$j(x,t) = -d\,^{RL}D_t^{1-\alpha}[grad\ c(x,t)], 0 < \alpha \leq 1. \tag{3}$$

It is clear that this non-locality is not unique and changes with respect to the diffused matter [9]. For instance, Compte and Metzler [10] considered long-time and short-time regimes of diffusion by modifying the Cattaneo's approach as follows

$$j(x,t) + \xi \frac{\partial^\alpha j(x,t)}{dt^\alpha} = -d\frac{\partial c(x,t)}{dx}, \tag{4}$$

where $\frac{\partial^\alpha}{dt^\alpha}$ denotes the Caputo fractional derivative which is the most used definition into the modelling of diffusion phenomena because the convenience of its mathematical description. However, since the power kernel function of the Caputo derivative is singular, it brings computational difficulties only removing by numerical techniques. Moreover, the fact that RL and Caputo fractional derivatives may be insufficient in the modeling physical processes that comply with exponential laws, such as heat diffusion, has led to the emergence of new generation fractional derivatives with regular kernels, namely Caputo-Fabrizio (CF) [11] and Atangana-Baleanu (AB) [12] derivatives. In this respect, Hristov proposed the heat diffusion equation with fading memory in terms of CF derivative by taking the $R(x,t)$ as Jeffrey's kernel function $R(t) = \exp[-(t-\tau)/\tau]$ (τ =constant relaxation time) into the Cattaneo's relation given by Eq.(2). This model is known as Cattaneo-Hristov (CH) model of elastic heat diffusion for which Koca and Atangana presented the solutions by applying Laplace transform and Crank-Nicholson schemes [13]. In addition, Sene gave a comparative research on the analytical solutions of classical, Caputo type and CH diffusion models [14]. Recently, Hristov has realized the similarity between the AB integral definition and Boltzman's superposition principle implying the fading memory which dates back 100 years. Thus, diffusion equation in terms of AB derivative in RL sense (ABR) has been constructed in [15]. Subsequently, Sene [16] has analyzed the analytical solutions of Hristov diffusion equations in an infinite domain by applying integral transform techniques.

The current chapter, which is motivated by the physical reality of Hristov's diffusion models, seeks to answer the question of how the ABR-type diffusion process behaves within a line segment under the effect of external sources. For this purpose, Fourier and Diethelm's predictor-corrector methods are combined to arrive the semi-analytical solutions. The results are supported by giving some illustrative examples under the variation of problem parameters.

2. MATHEMATICAL BACKGROUND

In this section, necessary mathematical tools for formulation of the problem can be reminded as follows.

Definition 1. *[17, 18] The RL fractional integral of a given function $f \in L_1(a,b)$ of order α is defined by*

$$^{RL}I_{a+}^{\alpha}f(t) = \frac{1}{\Gamma(\alpha)}\int_a^t (t-\tau)^{\alpha-1} f(\tau) d\tau, \quad \alpha > 0 \tag{5}$$

where $\Gamma(\alpha)$ denotes the Euler's gamma function.

Definition 2. *[19] The AB fractional integral operator is defined by*

$$^{AB}I_{a+}^{\alpha}f(t) = \frac{1-\alpha}{B(\alpha)}f(t) + \frac{\alpha}{B(\alpha)}{}^{RL}I_{a+}^{\alpha}f(t), \quad 0 < \alpha < 1, \tag{6}$$

in which $B(\alpha)$ denotes a smooth normalization function satisfying $B(0) = B(1) = 1$.

As in many studies in the literature, $B(\alpha) = 1$ is assumed for convenience in the diffusion problem discussed in this chapter.

Definition 3. *[19] The ABR derivative of a given function $f \in L_1(a,b)$ is defined by*

$$^{ABR}D_{a+}^{\alpha}f(t) = \frac{B(\alpha)}{1-\alpha}\frac{d}{dt}\int_a^t f(\tau) E_\alpha \left(-\frac{\alpha}{1-\alpha}(t-\tau)^\alpha\right) d\tau, \quad 0 < \alpha < 1 \tag{7}$$

Here, E_α denotes one parameter Mittag-Leffler function:

$$E_\alpha(z) = \sum_{k=0}^{\infty} \frac{z^k}{\Gamma(\alpha k + 1)}, \quad Re(\alpha) > 0, \ z \in \mathbb{C}. \tag{8}$$

Two parameter Mittag-Leffler function is also given by

$$E_{\alpha,\beta}(z) = \sum_{k=0}^{\infty} \frac{z^k}{\Gamma(\alpha k + \beta)}, \quad Re(\alpha) > 0, \ Re(\beta) > 0, \ z \in \mathbb{C}. \tag{9}$$

It is worth to note that $^{AB}I^{\alpha}_{a+}$ fractional integral is left and right inverse of the $^{ABR}D^{\alpha}_{a+}$ fractional differential operator, i.e.

$$^{ABR}D^{\alpha}_{a+}\left(^{AB}I^{\alpha}_{a+}f(t)\right) = f(t)$$

and

$$^{AB}I^{\alpha}_{a+}\left(^{ABR}D^{\alpha}_{a+}f(t)\right) = f(t).$$

The Hristov diffusion equation is based the constitutive mass balance equation giving a non-local relation between time ABR fractional derivative and concentration gradient as follows

$$^{ABR}D^{\alpha}_{a+}u(x,t) = -\frac{d}{dx}j_a(x,t). \tag{10}$$

On the other hand, there is an attractive mathematical similarity between the $^{AB}I^{\alpha}_{a+}$ definition and the Boltzman's linear superposition principle, namely

$$j(x,t) = m \nabla u(x,t) + \nu \int_0^t R(\xi,t) \nabla u(\xi,t) d\xi, \tag{11}$$

where m and ν correspond to transport coefficients depending on the prescribed diffusion process.

Boltzman's principle physically indicates that the current response of a material can be described by the superposition of the whole loading history of the material [20]. This physical concept is denoting the fading memory which is clearly explained by Coleman [21] as the distant past is less effective on the current values of the stress occurring into the material than the obtained deformations in the recent past.

By assuming $B(\alpha) = 1$ and also the transport coefficients as $m(\alpha) = 1 - \alpha$ and $\nu(\alpha) = \alpha$, the flux relaxation correlated the fading memory concept given by Eq.(11) is as follows

$$j_a(x,t) = {}^{AB}I^{\alpha}_{a+}\left(-d_0 \frac{\partial u}{\partial x}\right) \tag{12}$$

$$= -d_0 \left\{ m(\alpha) \frac{\partial u}{\partial x} + \nu(\alpha)^{AB}I^{\alpha}_{a+}\left(\frac{\partial u}{\partial x}\right)\right\}, \tag{13}$$

where d_0 is the diffusivity transport coefficient.

3. SEPARABLE SOLUTIONS VIA FOURIER METHOD

In this section, we investigate the separable solutions of the first type Hristov diffusion phenomena acting in a line segment $0 < x < \ell$. Let us consider the Hristov diffusion equation with source

$$^{ABR}D_{0+}^{\alpha}u(x,t) = d_0\,^{AB}I_{0+}^{\alpha}\frac{\partial^2 u(x,t)}{\partial x^2} + f(x,t), \quad (14)$$

in which d_0 denotes the diffusivity coefficient, under the homogeneous initial

$$u(x,0) = 0, \quad (15)$$

and the Dirichlet boundary

$$u(0,t) = u(\ell,t) = 0, \quad (16)$$

conditions. After adopting the relation between AB and RL fractional integrals, Eq.(14) can be rewritten as

$$^{ABR}D_{0+}^{\alpha}u(x,t) = d_0\left\{(1-\alpha)\frac{\partial^2 u(x,t)}{\partial x^2} + \alpha\,^{RL}I_{0+}^{\alpha}\frac{\partial^2 u(x,t)}{\partial x^2}\right\} + f(x,t). \quad (17)$$

To arrive the separable solutions, we first consider the homogeneous part of Eq.(17) and assume that

$$u(x,t) = U(x)T(t). \quad (18)$$

By substituting this separable solution into the homogeneous part

$$^{ABR}D_{0+}^{\alpha}u(x,t) = d_0\left\{(1-\alpha)\frac{\partial^2 u(x,t)}{\partial x^2} + \alpha\,^{RL}I_{0+}^{\alpha}\left[\frac{\partial^2 u(x,t)}{\partial x^2}\right]\right\}, \quad (19)$$

the main problem is reduced to two ordinary integer and fractional order differential equations:

$$\frac{d^2 U_k}{dx^2} + \lambda_k^2 U_k(x) = 0, \quad k = 1, 2, ..., \quad (20)$$

$$^{ABR}D_{0+}^{\alpha}T_k(t) = -d_0\lambda_k^2\left\{(1-\alpha)T_k(t) + \alpha\,^{RL}I_{0+}^{\alpha}T_k(t)\right\}, \quad (21)$$

where the eigenvalues $\lambda_k = \frac{k\pi}{\ell}$ the corresponding eigenfunctions $U_k(x) = \sin\left(\frac{k\pi x}{\ell}\right)$ are determined by using Eq.(20) and the boundary conditions $U(0) = U(\ell) = 0$. Therefore, we aim to find the series solution in the following form

$$u(x,t) = \sum_{k=1}^{\infty} \sin\left(\frac{k\pi x}{\ell}\right) T_k(t). \qquad (22)$$

To find the solution of non-homogeneous part, we assume the source function has the same eigenfunction expansion as

$$f(x,t) = \sum_{k=1}^{\infty} \sin\left(\frac{k\pi x}{\ell}\right) f_k(t), \qquad (23)$$

in which

$$f_k(t) = \frac{2}{\ell} \int_0^\ell f(\xi,t) \sin\left(\frac{k\pi \xi}{\ell}\right) d\xi. \qquad (24)$$

Next, substituting the series Eq.(22) and Eq.(23) into Eq.(17) and also applying the orthogonality property

$$\int_0^\ell \sin\left(\frac{k\pi x}{\ell}\right) \sin\left(\frac{p\pi x}{\ell}\right) dx = \begin{cases} \frac{\ell}{2}, & k=p \\ 0, & k \neq p \end{cases} \qquad (25)$$

we get

$$^{ABR}D_{0+}^\alpha T_k(t) = -d_0 \frac{k^2\pi^2}{\ell^2}\left\{(1-\alpha)T_k(t) + \alpha \, ^{RL}I_{0+}^\alpha T_k(t)\right\} + f_k(t), \qquad (26)$$

where the initial condition of the fractional order ordinary differential equation Eq.(26) is also calculated by substituting Eq.(15) into Eq.(22) as

$$T_k(0) = 0. \qquad (27)$$

Notice that the source function $f(x,t)$ specifies $f_k(t)$ and so the solution of $T_k(t)$. Therefore, different types of sources will be considered in the numerical procedure. Now, we will investigate the approximate solution belonging to fractional order initial value problem given by Eqs.(26)-(27).

4. NUMERICAL APPROXIMATION AND ILLUSTRATIVE EXAMPLES

The approximate solution of the reduced fractional order differential equation is based on the equivalence between the initial value problem Eqs.(26)-(27) and the following Volterra integral equation:

$$T_k(t) = -d_0 \frac{k^2\pi^2}{\ell^2} \left[(1-\alpha) {}^{AB}I_{0+}^\alpha T_k(t) + \alpha {}^{AB}I_{0+}^\alpha {}^{RL}I_{0+}^\alpha T_k(t) \right] + {}^{AB}I_{0+}^\alpha f_k(t). \quad (28)$$

In the above equation, we first substitute definition of AB integral into Eq.(28) and then rearrange the equation via unknown function $T_k(t)$ by taking into account the sequential integration property of RL integral [17]. Therefore, we get

$$T_k(t) = a_k \left[b_k {}^{RL}I_{0+}^\alpha T_k(t) + c_k {}^{RL}I_{0+}^{2\alpha} T_k(t) + (1-\alpha) f_k(t) + \alpha {}^{RL}I_{0+}^\alpha f_k(t) \right], \quad (29)$$

in which the coefficients are

$$a_k = \frac{1}{1 + d_0 \frac{k^2\pi^2}{\ell^2}(1-\alpha)^2},$$

$$b_k = -2d_0(1-\alpha)^2 \frac{k^2\pi^2}{\ell^2},$$

$$c_k = -d_0\alpha^2 \frac{k^2\pi^2}{\ell^2}.$$

After that, by using the following change of variables:

$$\tau_{k1}(t) = T_k(t),$$
$$\tau_{k2}(t) = {}^{RL}I_{0+}^\alpha T_k(t),$$

we transform the integral equation of order 2α to an integral equation system of order α which can be represented in the following matrix form

$$ {}^{RL}I_{0+}^\alpha \tau_k(t) = A_k \tau_k(t) + B_k(t), \quad \tau_k(0) = 0. \quad (30)$$

where

$$\tau_k(t) = \begin{bmatrix} \tau_{k1}(t) & \tau_{k2}(t) \end{bmatrix}^T,$$

$$A_k = \begin{bmatrix} 0 & 1 \\ \frac{1}{c_k} & -\frac{a_k b_k}{c_k} \end{bmatrix},$$

$$B_k(t) = \begin{bmatrix} 0 \\ -a_k \left[(1-\alpha) f_k(t) + \alpha {}^{RL}I_{0+}^\alpha f_k(t) \right] \end{bmatrix}.$$

To solve the Volterra integral equation system (30), we assume a finite time interval $[0, t_f]$ which is discretized into N equal parts with size of h labeled as t_i, $i = 0, 1, 2, ..., N$. Then, we use the Diethelm's predictor-corrector algorithm [23] to approximate the RL integral in Eq.30. So, we obtain the following discrete equation at node t_i as:

$$\sum_{j=0}^{i} \mu_{ij} T_k(t_j) = A_k T_k(t_i) + B_k(t_i), \quad T_k(t_0) = 0, \ i = 1, 2, ..., N, \quad (31)$$

where the μ_{ij} coefficients are determined as below

$$\mu_{ij} = \frac{h^\alpha}{\Gamma(\alpha+2)} \begin{cases} (i-1)^{\alpha+1} - i^{(\alpha+1)} + (\alpha+1)i^\alpha, & j=0, \\ (i-j+1)^{\alpha+1} + (i-j-1)^{\alpha+1} - 2(i-j)^{(\alpha+1)}, & 1 \leq j \leq i-1, \\ 1, & j=i. \end{cases} \quad (32)$$

Hence, the solution is achieved from Eq.(31) as

$$T_k(t_i) = (A_k - \mu_{ij}I)^{-1}\left[\sum_{j=0}^{i-1} \mu_{ij} T_k(t_j) - B_k(t_i)\right], \quad T_k(t_0) = 0, \ i = 1, 2, ..., N. \quad (33)$$

This numerical solution will be exhibited under different types of source functions $f(x,t)$ to show their effects on the Hristov diffusion. Therefore, we first evaluate $f_k(t)$ with respect to chosen functions $f(x,t)$ by using the relation Eq.(24). Then, we construct the time dependent matrices for the following cases.

- Case 1. $f(x,t) = x$:

  Since

  $$f_k(t) = (-1)^{k+1}\frac{2\ell}{k\pi} \quad \text{and} \quad {}^{RL}I^\alpha f_k(t) = (-1)^{k+1}\frac{2\ell}{k\pi}\frac{t^\alpha}{\Gamma(\alpha+1)},$$

  the time dependent matrix is

  $$B_k(t) = \begin{bmatrix} 0 \\ -a_k\left\{(-1)^{k+1}\frac{2\ell}{k\pi}\left(1 - \alpha + \frac{t^\alpha}{\Gamma(\alpha)}\right)\right\} \end{bmatrix}. \quad (34)$$

- Case 2. $f(x,t) = t$:

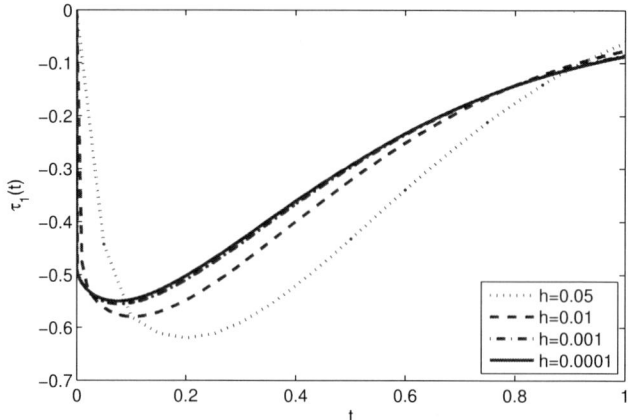

Figure 1. Dependence of the solution on the time discretizations for $\alpha = 0.7$ for Case 1.

Since
$$f_k(t) = t \text{ and } {}^{RL}I^\alpha f_k(t) = \frac{t^{\alpha+1}}{\Gamma(\alpha+2)},$$
the time dependent matrix is
$$B_k(t) = \begin{bmatrix} 0 \\ -a_k \left\{ (1-\alpha)t + \alpha \frac{t^{\alpha+1}}{\Gamma(\alpha+2)} \right\} \end{bmatrix}. \quad (35)$$

- **Case 3.** $f(x,t) = \sin\left(\frac{\pi x}{\ell}\right) e^{-t}$,

  Since
  $$f_k(t) = e^{-t} \text{ and } {}^{RL}I^\alpha f_k(t) = t^\alpha E_{1,1+\alpha}(-t)$$
  [22], the time dependent matrix is
  $$B_k(t) = \begin{bmatrix} 0 \\ -a_k \left\{ (1-\alpha)e^{-t} + \alpha t^\alpha E_{1,1+\alpha}(-t) \right\} \end{bmatrix}. \quad (36)$$

To depict each cases, we substitute the time dependent matrices Eqs.(34)-(36) into Eq.33, respectively. Firstly, the obtained solutions for Case 1 are exhibited to show the accuracy of the numerical solutions. For this purpose, Figure

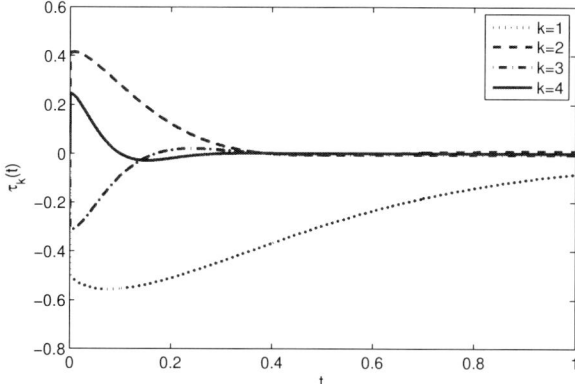

Figure 2. Contribution of eigencoordinates to the solution for $\alpha = 0.7$ and Case 1.

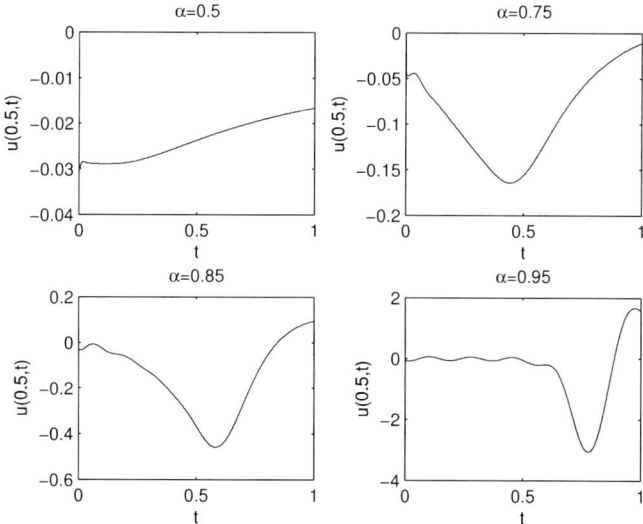

Figure 3. Effect of fractional order on the approximate solution of the diffusion equation.

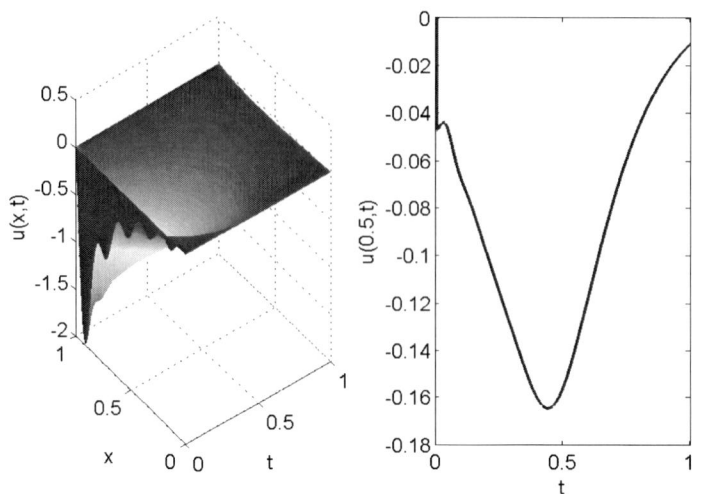

Figure 4. Approximate solution for Case 1 and $\alpha = 0.75$.

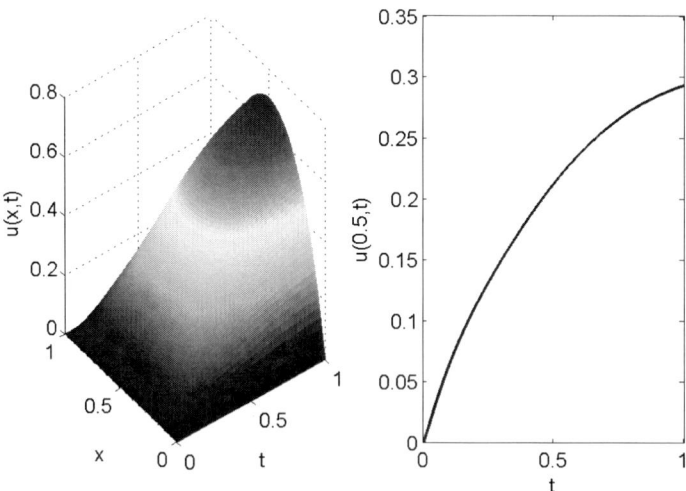

Figure 5. Approximate solution for Case 2 and $\alpha = 0.75$.

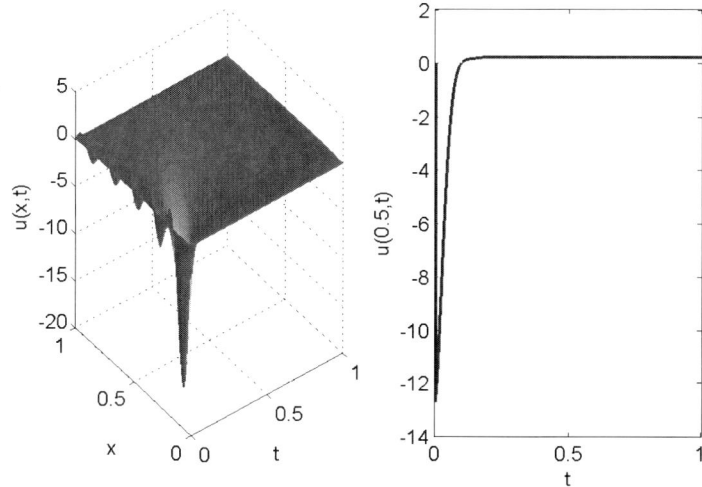

Figure 6. Approximate solution for Case 3 and $\alpha = 0.75$.

1 illustrates that the errors of the subsequent solutions tends to zero while the step size number is decreased. Secondly, Figure 2 shows that the series solution can be truncated since the contributions of the eigencoordinates decrease. It can be concluded from these figures that the numerical solution is convergent. Therefore, the rest of graphics is plotted for the step size $h = 0.001$ and the first 10 eigenfunctions. The diffusion solutions for changing values of α are depicted in Figure 3. According to this figure, solution curves evolve to wave behaviors and also the depth of waves increases when $\alpha \to 1$. Figure 4-6 are respectively exhibited for three cases according to $\alpha = 0.75$ which is intentionally fixed to clarify the effects of source functions. From these figures, we can understand that diffusion solutions strictly depend on the types of source functions.

CONCLUSION

This chapter has presented the semi-analytical solutions of the first type Hristov diffusion with a source in a line segment. Using the Fourier method, the eigenfunctions of the solution have been analytically achieved. Then, the reduced fractional order ordinary differential equation has been firstly converted to Volterra integral equation. To obtain the approximate solution, the integral equation of order 2α has been transformed to α order integral equation system by changing of variables. Utilizing the relation between AB and RL integrals, the integral equation system has been solved via Diethelm's predictor-corrector algorithm. The accuracy of the numerical solution has been graphically supported. In addition, dependence of the solutions on fractional parameter has been illustrated. It has been observed that different sources affect the depth of the behavior diffusion curves before relaxation.

REFERENCES

[1] Crank J., *The mathematics of diffusion*. Oxford University press, 1979.

[2] Cattaneo C., Sulla conduzione del calore. Atti *Sem. Mat. Fis. Univ. Modena* **3**, 83-101 (1948).

[3] Mokshin A.V., Yulmetyev R.M., & Hänggi P., Diffusion processes and memory effects, *New Journal of Physics* **7**(1), 9 (2005).

[4] Zheng L., Zhang X., *Modeling and analysis of modern fluid problems*. Academic Press, 2017.

[5] Mainardi F., Gorenflo R., Time-fractional derivatives in relaxation processes: a tutorial survey, *Fractional Calculus and Applied Analysis* **10**(3), 269-308 (2007).

[6] Metzler R., Klafter J., The random walk's guide to anomalous diffusion: a fractional dynamics approach, *Physics Reports* **339**(1), 1-77 (2000).

[7] Metzler R., Klafter J., The restaurant at the end of the random walk: recent developments in the description of anomalous transport by fractional dynamics, *Journal of Physics A: Mathematical and General* **37**(31), R161 (2004).

[8] Povstenko Y., *Linear fractional diffusion-wave equation for scientists and engineers*. Springer International Publishing, Switzerland, 2015.

[9] Ajlouni A.W.M., Al-Rabai'ah H.A., Fractional-calculus diffusion equation. *Nonlinear Biomedical Physics* **4**(1), 3 (2010).

[10] Compte A., Metzler R., The generalized Cattaneo equation for the description of anomalous transport processes, *Journal of Physics A: Mathematical and General* **30**, 7277e7289 (1997).

[11] Caputo M., Fabrizio M., A new definition of fractional derivative without singular kernel, *Progr. Fract. Differ. Appl.* **1**(2), 1-13 (2015).

[12] Atangana A., Baleanu D., New Fractional Derivatives with Non-local and Non-Singular Kernel Theory and Application to Heat Transfer Model, *Thermal Science* **20**(2), 763-769 (2016).

[13] Koca I., Atangana A., Solutions of Cattaneo-Hristov model of elastic heat diffusion with Caputo-Fabrizio and Atangana-Baleanu fractional derivatives, *Thermal Science* **21**(6 Part A), 2299-2305 (2017).

[14] Sene N., Solutions of fractional diffusion equations and Cattaneo-Hristov diffusion model. *International Journal of Analysis and Applications* **17**(2), 191-207 (2019).

[15] Hristov J., On the Atangana–Baleanu derivative and its relation to the fading memory concept: the diffusion equation formulation. In Fractional Derivatives with Mittag-Leffler Kernel (pp. 175-193). Springer, Cham. (2019).

[16] Sene N., Analytical solutions of Hristov diffusion equations with non-singular fractional derivatives, *Chaos: An Interdisciplinary Journal of Nonlinear Science* **29**(2), 023112 (2019).

[17] Samko S.G., Kilbas A.A., Marichev O.I., *Fractional integrals and derivatives (Vol. 1)*. Gordon and Breach Science Publishers, Switzerland 1993.

[18] Podlubny I., *Fractional differential equations: an introduction to fractional derivatives, fractional differential equations, to methods of their solution and some of their applications*. Elsevier. (1998).

[19] Baleanu D., Fernandez A., On some new properties of fractional derivatives with Mittag-Leffler kernel, *Communications in Nonlinear Science and Numerical Simulation* **59**, 444-462 (2018).

[20] Malkin A.Y., Isayev A.I., *Rheology: concepts, methods, and applications*. Elsevier (2017).

[21] Coleman B.D., Gurtin M.E., Equipresence and constitutive equations for rigid heat conductors. *Zeitschrift für angewandte Mathematik und Physik ZAMP* **18**(2), 199-208 (1967).

[22] Garrappa R., Kaslik E., Popolizio M., Evaluation of fractional integrals and derivatives of elementary functions: Overview and Tutorials, *MDPI Mathematics* **7**(5), 407(21 pages) (2019).

[23] Diethelm K., Ford N.J., Freed A.D., A predictor-corrector approach for the numerical solution of fractional differential equations, *Nonlinear Dynamics* **29**(1-4), 3-22 (2002).

In: A Closer Look at the Diffusion Equation ISBN: 978-1-53618-330-6
Editor: Jordan Hristov © 2020 Nova Science Publishers, Inc.

Chapter 6

NON-GAUSSIAN DIFFUSION EMERGENCE IN SUPERSTATISTICS

Maike A. F. dos Santos[1,2,*]
[1]Department of Physics, Pontifical Catholic University,
Rio de Janeiro, RJ, Brazil
[2]Institute of Science and Technology for Complex Systems,
Rio de Janeiro, RJ, Brazil

Abstract

An increasing number of researches have reported non-Gaussian diffusion process that has a linear mean square displacement in time, i.e. $\langle (x - \langle x \rangle)^2 \rangle \sim t$, also known as "non-Gaussian yet Brownian". Such processes are atypical within diffusive processes. Thus, non-Gaussian yet Brownian has been reported in complex systems associated with biological, active, and soft matter systems. In this broad scenario, this chapter shows how we can carry out a superstatistical approach of Brownian diffusive processes associated with fluctuation in diffusivities to build overall distributions connected with non-Gaussian diffusion. In this sense, we initially approach a simple and recurring case in the literature that is associated with the exponential distribution for diffusivities and how such process generates a Laplace diffusion. Besides, the chapter covers general distributions associated with three-parameter Mittag-Leffler distribution to diffusivities which implies a new and broad class of overall

---

[*]Corresponding Author's Email: santosmaikeaf@gmail.com.

distributions linked to non-Gaussian diffusion. Finally, the chapter develops superstatistics for a version of the Fokker-Planck equation to address system that considers fluctuations in diffusivity and a linear force in position, i.e., $F(x) \propto -x$. The theoretical tools presented in this chapter aim to provide the reader with the main characteristics of the non-Gaussian diffusion.

Keywords: diffusion equation, non-Gaussian diffusion, generalised distributions, superstatistics

1. INTRODUCTION

Nowadays it is well known that Gaussian diffusion is closely connected to the diffusion of free particles immersed in a homogeneous medium in thermodynamic equilibrium, that from the statistical point of view correspond to identically distributed, random variables (the steps of the random walk) with finite variance [1]. The beginning of this type of diffusion came with experiments carried out in 1828 by the botanist Robert Brown, who reported the incessant and irregular movement of tiny particles immersed in water [2]. This type of random movement became known as Brownian motion and is intimately connected to the thermal agitation of the atoms and molecules in the medium. The appropriate mathematical description was proposed only in 1905 by Einstein [3] using the diffusion equation $\partial_t P(x,t) = D_0 \partial_x^2 P(x,t)$, where D_0 is the diffusivity of the medium and $P(x,t)$ is a probability distribution function. In this context, the Brownian movement is entirely characterised by two properties. The first one is related to the fact that the mean square displacement (MSD) has a linear relation with time, i.e., $\langle (x - \langle x \rangle)^2 \rangle \propto t$. The second is associated with the Gaussian distribution emergence, as follows

$$P(x,t) = \frac{\exp\left[-\frac{x^2}{4D_0 t}\right]}{\sqrt{4\pi D_0 t}}, \qquad (1)$$

in which $x \in \mathbb{R}$ and the initial condition is given by $\lim_{t \to 0} P(x,t) = \delta(x)$, being $P(x,t)$ the probability of finding the particle at a position x at time t. After Einstein's proposal [3], a series of important works validated the diffusion today known as Brownian, among them the Perrin J.'s research had a fundamental role [4]. Others theoretical advances in diffusion context came with the Fokker-Planck equation, which is a version of the diffusion equation that includes a force applied under the system [1]. In this context, diffusion with mean square displacement is linear at time with Gaussian distribution is strongly associated with Boltzmann-Gibbs statistics [1].

A more general perspective on diffusive processes has emerged with a series of experiments [5, 6] reporting anomalous diffusive processes, i.e., $\langle x^2 \rangle \propto t^\delta$ with $\delta \neq 1$, accompanied by non-Gaussian distributions. In this direction, several generalised diffusive models have been proposed to describe the anomalous diffusion process, among the most successful we can include: the fractional [5, 6], heterogeneous [7, 8, 9] and non-linear [10] approach to the diffusion equation. After the beginning of this century, with many experimental studies associated with the diffusion of tiny particles in complex systems, there was a need for more general approaches to statistics associated with Brownian diffusion. In this sense, Beck C. and E. Cohen proposed a generalisation of Boltzmann-Gibbs statistics [11, 12], which was known as *superstatistics*. Beck-Cohen's theory considers a Langevin equation associated with an incomplete *"ensemble"* (with imperfections) through a conditional probability that admits an additional statistical treatment on an intensive parameter that fluctuates, an example of intensive parameters that fluctuate and may describe by superstatistics include temperature, diffusivity and mass. Thus, superstatistics consists of a statistic from another statistic. In this sense, considering a statistic about an $f(\beta)$ distribution associated with Boltzmann and $e^{-\beta E}$ factors, Tsallis et al. constructed a general entropy to Beck-Cohen superstatistics [13]. In particular, the superstatistics of Boltzmann factors with χ^2-distribution implies Tsallis statistical mechanics [11, 12, 13, 14]. Thus, Cohen-Beck's theory gained notoriety in stochastic treatment in complex systems based on experimental data ranging from turbulent processes [15] to air pollution statistics [16]. Immersed in the idea of superstatistics, Wang B. *et al.* showed that colloidal beads on phospholipid bilayer and beads diffuse through pores in the entangled F-actin networks have a non-Gaussian diffusion with linear MSD over time, i.e., MSD$\sim t$, this type of diffusion came to be called "non -Gaussian yet Brownian" [17, 18]. In this scenario, a common example of non-Gaussian diffusion is given by Laplace-*like* distribution as follow

$$P(x,t) = \frac{\exp\left[-\frac{|x|}{\lambda(t)}\right]}{2\lambda(t)}, \qquad (2)$$

in which $x \in \mathbb{R}$. In the case when $\lambda(t) = \sqrt{Dt}$, we have the exponential distribution for the diffusivities reported in Ref. [17]. This type of diffusion has been reported in biological, soft, and active matter systems [19, 20, 21, 22, 23, 24, 25, 26, 27, 28, 29].

In this broad context, this chapter aims to lead the reader to understand how to make statistics of Gaussian process that imply non-Gaussian diffusion. More specifically, in the second section of this chapter, we will present how the idea of a series of Brownian particles associated with different diffusivities can be determined addressed by superstatistics. In this direction, in section (2) we address a special cases to the reader, showing how an exponential distribution for the diffusivities of a system implies Laplace diffusion with MSD$\sim t$. Besides, we introduced the Mittag-Leffler function with three parameters to construct a $p(D)$ distribution of diffusivities that is normalizable and that recovers the exponential case. In this sense, we will present a new class of non-Gaussian diffusion. Thus, we show particular situations that recover the result found by R. Jain et al. [30]. In the third section, we consider the Fokker-Planck equation with a fluctuating diffusivity parameter D and a harmonic potential, i.e., $0 \leq V(x) \sim x^2$, associated with a stationary distribution of diffusivities $p(D)$ constructed in second section. In this sense, this chapter provides ways to understand the non-Gaussian diffusion process in superstatistics

2. Non-Gaussian Diffusion and Superstatistics

Motivated by the diversity of diffusion processes that present non-Gaussian diffusion, a superstatistics on overall distribution function $P(x,t)$ of a system of particles that moves individually on different D_i domains (see Fig. 1). Then, the junction patches with local diffusivity D becomes the weighted average, where $p(D)$ is the stationary probability density for diffusivities D.

Now, we can introduce the overall average under Gaussian process and their probabilistic-diffusivity weight as follows [11, 12, 18, 19]

$$P(x,t) = \int_0^\infty p(D) f(x,t|D) dD, \quad \left(f(x,t|D) = \frac{\exp\left[-\frac{x^2}{4Dt}\right]}{\sqrt{4\pi Dt}} \right) \quad (3)$$

in which $p_D = p(D)$ ($0 \leq D \leq \infty$) is the weighted average of diffusivities

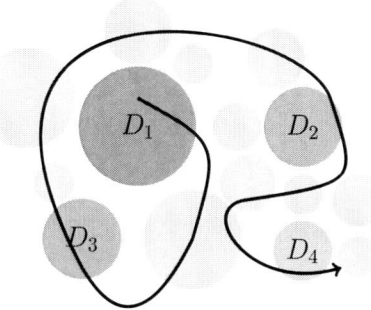

Figure 1. This figure shows the random environment with different diffusivity patches.

and the conditional probability $f(x,t|D)$ given by Gaussian distributions (between parentheses) associate with Brownian process. The overall distribution $P(x,t)$ assume a Gaussian form (Eq. 1)) when a particular diffusion domain D_0 overlaps all the others diffusivities, its presented by a collapsed diffusivity density in D_0 value, i.e., $p(D) = \delta(D - D_0)$. For a non collapsed density diffusivity we have a system with non-equilibrium stationary state [19]. In the general context, we consider the Fourier transform $\mathcal{F}\{P(x,t)\} = \overline{P}(k,t) = \int_{-\infty}^{\infty} P(x,t)e^{-ixk}dx$ in Eq. (3) that implies

$$\begin{aligned}\overline{P}(k,t) &= \int_0^{\infty} p(D)e^{-Dtk^2}dD \\ &= \widetilde{p}(tk^2),\end{aligned} \quad (4)$$

in which $\widetilde{g}(s) = \int_0^{\infty} e^{-st}g(t)dt$ is the Laplace transform. The Eq. (4) were found in Ref. [19]. Thereby the Fourier inverse transform of the Eq. (4) is rewrite by

$$\begin{aligned}P(x,t) &= \frac{1}{2\pi t^{\frac{1}{2}}}\int_{-\infty}^{\infty} e^{ix\kappa/t^{\frac{1}{2}}}\widetilde{p}(\kappa^2)d\kappa, \\ &= \frac{F[|\zeta|]}{t^{\frac{1}{2}}},\end{aligned} \quad (5)$$

in which $\zeta = x/t^{\frac{1}{2}}$ and $F[\zeta] = \pi^{-1}\int_0^{\infty}\cos[\zeta\kappa]\widetilde{p}(\kappa^2)d\kappa$. The Eq. (5) revels a power-law function to return probability $P(x=0.t) \sim t^{-\frac{1}{2}}$. A typical distribution discussed in contexts of biological and soft matter systems [17, 18, 19] is the exponential distribution

$$p(D) = \frac{\exp\left[-\dfrac{D}{D_0}\right]}{D_0},\qquad(6)$$

in which $D \in \mathbb{R}^+$. Using the Eq. (6) in Eq. (3) obtains

$$P(x,t) = \frac{\exp\left[-\dfrac{|x|}{\sqrt{tD_0}}\right]}{2\sqrt{tD_0}},\qquad(7)$$

that is non-Gaussian form associate to Laplace distribution. However, the diffusion process is yet Brownian

$$\langle x^2 \rangle = 2D_0 t,\qquad(8)$$

in which D_0 is the effective diffusivity. In this sense, we can introduce a more general distribution based on the three-parameter Mittag-Leffler function as follows

$$p(D) = \mathcal{N} D^{\beta-1} E_{\alpha,\beta}^{\delta}\left[-\frac{D^\alpha}{D_0}\right],\qquad(9)$$

in which $\mathcal{N}$ is normalisation constant and $E_{\alpha,\beta}^{\delta}[z]$ is the three parameter Mittag-Leffler function is defined by [31]

$$E_{\alpha,\beta}^{\delta}[z] = \sum_{n=0}^{\infty} \frac{(\delta)_n}{\beta[\alpha n + \beta]} \frac{z^k}{k!},\qquad(10)$$

in which $(\delta)_n = \Gamma[n+\delta]/\Gamma[\delta]$ and $0 < \alpha, 0 < \beta$ and $0 < \delta$. To $\alpha = \beta = \delta = 1$ we have $E_{1,1}^{1}[z] = e^z$ that implies exponential distribution of diffusivity. The Mittag-Leffler function appears in many systems associate to fractional calculus applications in diffusion process [32, 33, 34, 35, 36, 37, 38, 39, 40, 41, 42]. For $\alpha = 1, \beta = \delta$ we have $p(D) = \mathcal{N} D^{\delta-1} \exp[-D/D_0]$ that for $\delta = n/2$ recover the result reported in Ref. [30] that consider n as a dimensional number and D is the modulus of a n-dimensional vector.

Is well known the Laplace transform form of three parameter Mittag-Leffler function as

$$\mathcal{L}\{E_{\alpha,\beta}^{\delta}[-\nu t^\alpha]\} = s^{\alpha\delta-\beta}(s^\alpha + \nu)^{-\delta}.\qquad(11)$$

Thereby, applying the Laplace transform in Eq. (9) we obtain

$$\widetilde{p}(tk^2) = \frac{\mathcal{N}_{\alpha,\beta}^{\delta}(t|k|^2)^{\alpha\delta-\beta}}{(D_0^{-1} + (t|k|^2)^{\alpha})^{\delta}}, \tag{12}$$

with follow normalisation limit

$$\begin{aligned}\mathcal{N} &= \frac{1}{\lim_{s \to 0} \widetilde{p}(s)} \\ &= \lim_{s \to 0} s^{\beta-\alpha\delta}(D_0^{-1} + s^{\alpha})^{\delta}. \end{aligned} \tag{13}$$

Then $p(D)$ is normalizable to $\alpha\delta = \beta$ with $0 < \alpha$ and $\delta \in \mathcal{R}^+$ that implies $\mathcal{N} = D_0^{-\delta}$. Thereby, we has the follow distribution

$$p(D) = \frac{D^{\alpha\delta-1}}{D_0^{\delta}} E_{\alpha,\alpha\delta}^{\delta}\left[-\frac{D^{\alpha}}{D_0}\right]. \tag{14}$$

However, the distribution (14) not implies a finite average of D for $\alpha < 1$, i.e., $\langle D \rangle_{p(D)} \to \infty$. The Mittag-Leffler density distribution (14) has the similar aspects with superstatistics proposed in Ref. [43] to approach generalised Maxwell distributions. Using the Eq. (14) in Eq. (4) we obtain

$$\overline{P}(k,t) = \frac{D_0^{-\delta}}{(D_0^{-1} + (t|k|^2)^{\alpha})^{\delta}}, \tag{15}$$

that have two important limits, $\alpha \to 1$ with $0 < \delta$ and $\delta \to 1$ with $0 < \alpha < 1$. To the first one, $\alpha \to 1$, the inverse Fourier transform of the overall distribution is given by

$$\begin{aligned}P_\delta(x,t) &= \lim_{\alpha \to 1} \mathcal{F}^{-1}\{\widetilde{p}(tk^2)\} \\ &= \left(\frac{|x|}{\sqrt{D_0 t}}\right)^{\delta-\frac{1}{2}} \frac{2^{\frac{1}{2}-\delta} K_{\frac{1}{2}-\delta}\left[\frac{|x|}{\sqrt{D_0 t}}\right]}{(tD_0)^{\frac{1}{2}}\sqrt{\pi}\Gamma[\delta]}, \end{aligned} \tag{16}$$

in which $K_a[z]$ is a hyperbolic Bessel functions of second kind. This distribution implies the second moment as follows

$$\langle x^2 \rangle = 2D_0 \delta t, \tag{17}$$

to $\delta = 1$ we recover the non-Gaussian diffusion reported in Refs. [17, 18] and for $\delta = n/2$ we recover the R. Jain *et al.* result [30]. The distribution (16) is exemplified in Fig. (2-a) for different δ values.

The second limit is $\delta \to 1$ in Eq. (15) is exemplified in Fig. (2-b) for different α values between $1/2 < \alpha < 1$. For $\alpha < 1$ we have $\langle x^2 \rangle \to \infty$ and for that reason, we can call the case $\alpha < 1$ to $\delta \in \mathrm{R}^+$ as Lévy-*like* diffusion. An interesting case about the distribution (15) is the parameters that make it non-normalizable. An example of this occurs for $\alpha\delta > \beta$ with $0 < \alpha, 0 < \beta$ and $\delta \in \mathrm{R}^+$. The non-normalised distribution function in Physics has been a great challenge to current research topics [44, 45].

A series of other overall distribution Eq. (3) have been defined in the literature from different distributions $p(D)$. Particularly, important cases were addressed by V Sposini *et al.* which considers a heterogeneous ensemble of Brownian particles (HEBP) with the distribution of diffusivities that generate overall distribution with stretched exponential forms and Lévy distribution [46, 47]. The χ^2-gamma distribution was considered by S Hapca *et al.* to describe the anomalous diffusion of heterogeneous populations characterised by normal diffusion at the individual level [48]. In this broad scenario, a series of non-Gaussian processes can be built to capture the essence of different diffusive processes. A question that can be considered here and whether there is another quantity that has an invariant behaviour in a mathematical form analogous to the linear relation between second moment and time. The answer is simple when we look at entropy associated to the overall distribution (5) we obtain [47]

$$\frac{S(t)}{k_B} = -\int_{-\infty}^{\infty} P(x,t) \ln P(x,t) = A + \frac{1}{2} \ln t, \qquad (18)$$

that is a universal behaviour of Eq. (5). This class system takes into account large deviations of diffusivities of systems in non-equilibrium stationary states [47].

3. SUPER FOKKER-PLANCK EQUATION FOR THE HARMONIC POTENTIAL

We can move forward in our analysis by considering the Fokker-Planck (FP) equation in the context of superstatistics. In this scenario, the FP equation now has diffusivity D associated with a density distribution $p(D)$. A similar model

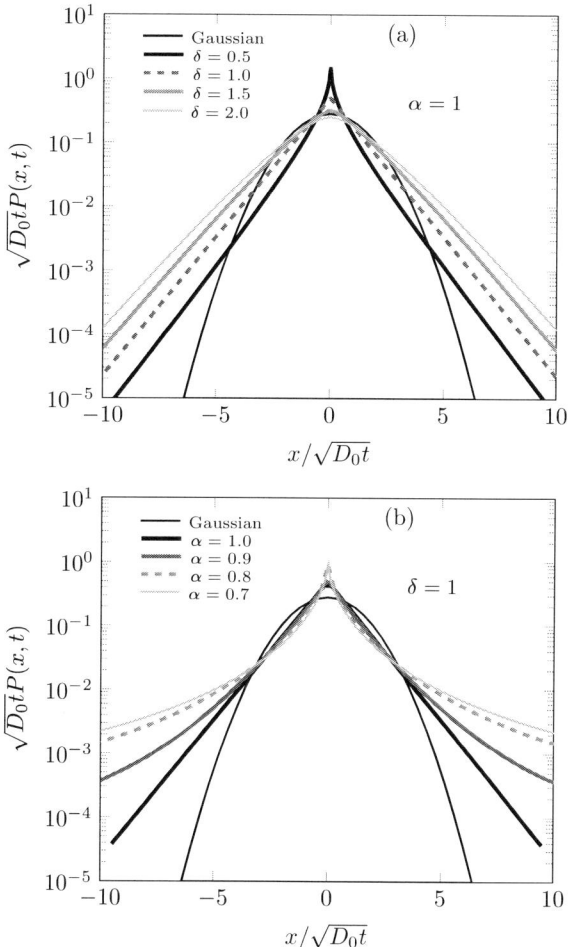

Figure 2. The Fig. (a) shows non-Gaussian distributions to different values associate to Eq. (16). The Fig. (b) shows non-Gaussian distributions to different α values associate to the inverse Fourier transform of Eq. (15) with $\delta = 1$. In both figures, the blue curve represents the Gaussian diffusion, i.e., Eq. (1).

with fluctuations on inverse diffusivity $\beta = 1/D$ was considered by A A Budini *et al.* in Ref. [49]. So, here we call super FP equation the following equation

$$\frac{\partial}{\partial t}f(x,t|D) = \frac{\partial}{\partial x}\left\{D\frac{\partial}{\partial x}f(x,t|D) + \frac{c}{2}xf(x,t|D)\right\}, \quad (19)$$

with $0 \leq c$ with $f(x, 0|D) = \delta(x)$ in which $f(x, t|D)$ is a probability to find the particle on x position at time t in a random packet with indexed diffusivity D (see Fig. (1)). When we mix and saute the heterogeneous ensemble of Brownian particles by uses of the Eq. (3) and conditional probability $f(x, t|D)$, we obtain the overall probability of finding the particle on position x at time t. Its quantity is observable in single-particle tracking that permit understands the trajectories of a single molecule in biological systems [50]. Now, to solve the Eq. (19) we apply Fourier transform, that implies

$$\frac{\partial}{\partial t}\overline{f}(k,t|D) = -k^2 D\overline{f}(k,t|D) - \frac{c}{2}k\frac{\partial}{\partial k}\overline{f}(k,t|D), \tag{20}$$

with $\overline{f}(k, 0|D) = 1$. So, we consider the *ansatz* as a solution in Fourier space as follows

$$\overline{f}(k,t|D) = \exp[-a_t k^2], \tag{21}$$

in which $a_t = a(t)$ is a function that we need to determine. Applying Eq. (21) in Eq. (20) we have

$$\frac{d}{dt}a(t) = D - ca(t), \tag{22}$$

in which $a(0) = 0$ we have $\overline{f}(k,0) = 1$. Thereby, the Eq. (22) implies

$$a(t) = \frac{D}{c}(1 - e^{-ct}), \tag{23}$$

that in ansatz (21) make complete the solution (21) in Fourier space. We have followed solution

$$\begin{aligned} f(x,t|D) &= \mathcal{F}^{-1}\left\{\exp\left[-\frac{D}{c}(1-e^{-ct})k^2\right]\right\} \\ &= \frac{1}{\sqrt{4\pi D c^{-1}(1-e^{-ct})}}\exp\left[-\frac{cx^2}{4D(1-e^{-ct})}\right]. \end{aligned} \tag{24}$$

This result permit the use of Eq. (3) determine the overall probability. Thereby, we can write the overall probability as a function of diffusivity distribution $p(D)$ in Laplace space as follows

$$\begin{aligned} \overline{P}(k,t) &= \int_0^\infty p(D)\exp\left[-\frac{D}{c}(1-e^{-ct})k^2\right]dD \\ &= \widetilde{p}\left[c^{-1}(1-e^{-ct})k^2\right]. \end{aligned} \tag{25}$$

Non-Gaussian Diffusion Emergence in Superstatistics

Based on previous analyses we consider three cases to analyse. The exponential distribution to diffusivity and the Mittag-Leffler distribution (for two limits). So, the exponential distribution to $p(D)$ (Eq. (6)) in Eq. (25) implies

$$P(x,t) = \frac{\exp\left[-\frac{|x|}{\sqrt{(1-e^{-ct})(D_0/c)}}\right]}{2\sqrt{(1-e^{-ct})(D_0/c)}}, \quad (26)$$

that for $t \to \infty$ we obtain a Laplace distribution to stationary state. This stationary state was found in stochastic resetting problem [37].

Now, we can consider the normalised diffusivity density present in Eq. (14), that in overall probability (3) implies in a distribution with two parameters, α and δ, that in Fourier space is write as follows

$$\overline{P}(k,t) = \left(1 + \left((1-e^{-ct})\left(D_0^{\frac{1}{\alpha}}/c\right)|k|^2\right)^\alpha\right)^{-\delta}. \quad (27)$$

To understand the influence of (α, δ) indexes we can approach both cases separately. So, the function $P_\delta(x,t) = \lim_{\alpha \to 1} \mathcal{F}^{-1}\{\widetilde{p}(tk^2)\}$ is write as follows

$$P_\delta(x,t) = \left(\frac{|x|}{\sqrt{D_0 c^{-1}(1-e^{-ct})}}\right)^{\delta-\frac{1}{2}} \frac{2^{\frac{1}{2}-\delta} K_{\frac{1}{2}-\delta}\left[\frac{|x|}{\sqrt{D_0 c^{-1}(1-e^{-ct})}}\right]}{(D_0 c^{-1}(1-e^{-ct}))^{\frac{1}{2}}\sqrt{\pi}\Gamma[\delta]}, \quad (28)$$

to $\delta \to 1$ we recover the solution (26). The second moment of Eq. (28) is given by

$$\langle x^2 \rangle = \frac{2\delta D_0}{c}(1-e^{-ct}), \quad (29)$$

that is $\langle x^2 \rangle \sim 2\delta D_0 t$ to short times and $\langle x^2 \rangle \sim 2\delta D_0/c$ to long times. Moreover, the stationary solution of Eq. (28) is given by

$$P_\delta^{stat}(x) = \left(\frac{|x|}{\sqrt{D_0 c^{-1}}}\right)^{\delta-\frac{1}{2}} \frac{2^{\frac{1}{2}-\delta} K_{\frac{1}{2}-\delta}\left[\frac{|x|}{\sqrt{D_0 c^{-1}}}\right]}{c^{-\frac{1}{2}} D_0^{\frac{1}{2}}\sqrt{\pi}\Gamma[\delta]}. \quad (30)$$

Different cases associated with Eq. (28) are presented in Fig. (3-a) to exemplify the temporal evolution. The curves in the figure were compared with the

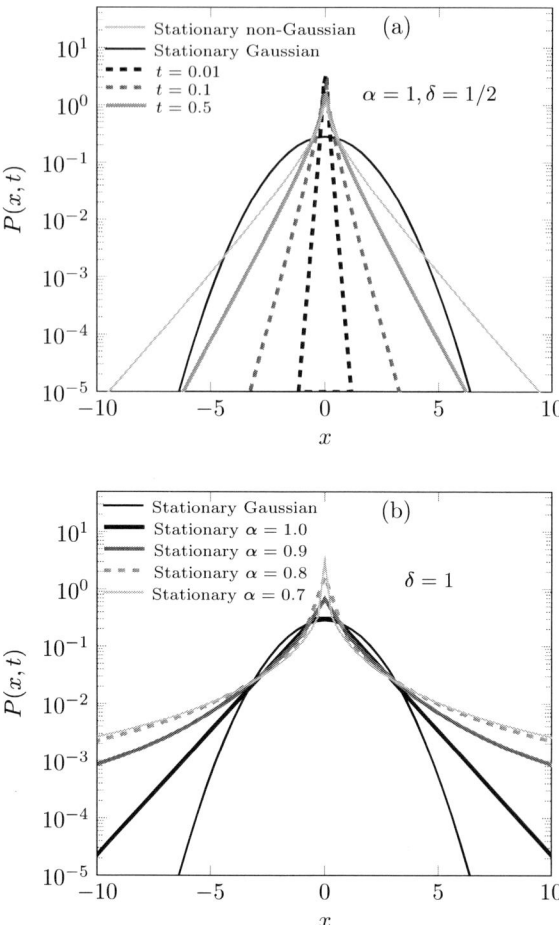

Figure 3. The Fig. (a) show non-Gaussian distributions considering $\delta = 1/2$, $c = 1$, $D_0 = 1$ and different times in Eq. (28). The Fig. (b) show non-Gaussian distributions considering $\delta = 1$ $c = 1$, $D_0 = 1$ and different α values associate to inverse Fourier transform of Eq. (27) in stationary state. In both figures the blue curve represent the overall distribution (3) of conditional probability (24) with $p(D) = \delta(D - D_0)$, $c = 1$ and $D_0 = 1$.

Gaussian distribution in the stationary case, i.e., $p(D) = \delta(D - D_0)$ in Eq. (24) for long times. Besides, the stationary solution (30) shows a similarity with the result of C Beck *et al.* [51] who investigated connections between super ensembles in systems with position-dependent diffusivity.

Another important situation is given by limit $\delta \to 1$ that can be obtained with the inverse numerical Fourier transform in Eq. (27). Thus, in Fig. (3-b) we consider $\delta = 1$ to present the stationary solution for different values of α. In this context, we obtain generalised Lévy-*like* distributions, as the second moment does not exist for $\alpha < 1$, i.e., $\langle x^2 \rangle \sim \infty$, which is a typical characteristic associate to Lévy flights [6]. However, in the case addressed here, the return probability is given by

$$P(0,t) \sim t^{-\frac{1}{2}}. \tag{31}$$

This result differs from the return probability of Lévy flights [38], which is $P(0,t) \sim t^{-\frac{1}{\mu}}$.

A broader context was considered by M. V. Chubynsky and G. W. Slater [52] who considered situations in which diffusivities are not associated with a stationary distribution, but with a process that evolves over time, i.e., $p(D,t)$. Such processes came to be called "diffusing diffusivity" and important contributions have been explored in a series of recent works [19, 20, 30, 46, 47].

CONCLUSION

The chapter demonstrated the development of non-Gaussian diffusion through superstatistics approach. We show that these techniques include the mixture of the heterogeneous ensemble of Brownian particles approached by a definition of an overall distribution associate to non-Gaussian diffusion. The chapter presents the connection between the exponential distribution of diffusivities and Laplace diffusion. In the sequence, the chapter presented the detailed construction of a diffusivity distribution through Mittag-Leffler function with two parameters and as such development implies a rich class of non-Gaussian diffusion that generate the diffusive processes that are "*yet Brownian*" for $\alpha = 1$ and Lévy-*like* for $0 < \alpha < 1$.

REFERENCES

[1] Risken H., Fokker-Planck equation. In: *The Fokker-Planck Equation*, Springer, Berlin, Heidelberg, (1996).

[2] Brown R., A brief account of microscopical observations made in the months of June, July and August 1827, on the particles contained in the pollen of plants; and on the general existence of active molecules in organic and inorganic bodies, *The Philosophical Magazine*, **4**, 161-173 (1828).

[3] Einstein A., On the motion of small particles suspended in liquids at rest required by the molecular-kinetic theory of heat, *Annalen der physik*, **17**, 208 (1905).

[4] Perrin J., Comptes rendus, *Ann. Chim. Phys*, **18**, (1909).

[5] Metzler R. and Klafter J., The random walk's guide to anomalous diffusion: a fractional dynamics approach, *Physics reports*, **339**, 1-77 (2000).

[6] dos Santos M. A. F., Analytic approaches of the anomalous diffusion: A review. *Chaos, Solitons and Fractals*, **124**, 86-96 (2019).

[7] Itô K., Stochastic integral. *Proceedings of the Imperial Academy*, **20**, 519-524 (1944).

[8] Stratonovich R. L., A new representation for stochastic integrals and equations, *SIAM Journal on Control*, **4**, 362-371 (1966)

[9] Hanggi P., Nonlinear fluctuations: The problem of deterministic limit and reconstruction of stochastic dynamics, *Physical Review A*, **25**, 1130 (1982).

[10] Frank T. D., *Nonlinear Fokker-Planck equations: fundamentals and applications*, Springer Science and Business Media, (2005).

[11] Beck C., Dynamical foundations of nonextensive statistical mechanics, *Physical Review Letters*, **87**, 180601 (2001).

[12] Beck C. and Cohen E. G., Superstatistics. *Physica A: Statistical mechanics and its applications*, **322**, 267-275 (2003).

[13] Tsallis C. and Souza A. M., Constructing a statistical mechanics for Beck-Cohen superstatistics, *Physical Review E*, **67**, 026106 (2003).

[14] Tsallis C., Possible generalization of Boltzmann-Gibbs statistics, *Journal of statistical physics*, **52**, 479-487 (1988).

[15] Beck C., Superstatistics in hydrodynamic turbulence, *Physica D: Nonlinear Phenomena*, **193**, 195-207 (2004).

[16] Williams G., Schäfer B. and Beck C., Superstatistical approach to air pollution statistics, *Physical Review Research*, **2**, 013019 (2020).

[17] Wang B., Anthony S. M., Bae S. C. and Granick S., Anomalous yet brownian, *Proceedings of the National Academy of Sciences*, **106**, 15160-15164 (2009).

[18] Wang B., Kuo J., Bae S. C. and Granick S., When Brownian diffusion is not Gaussian, *Nature materials*, **11**, 481-485 (2012).

[19] Chechkin A. V., Seno F., Metzler R. and Sokolov I. M., Brownian yet non-Gaussian diffusion: from superstatistics to subordination of diffusing diffusivities, *Physical Review X*, **7**, 021002 (2017).

[20] Metzler R., Superstatistics and non-Gaussian diffusion, *The European Physical Journal Special Topics*, **229**, 711-728 (2020).

[21] Toyota T., Head D. A., Schmidt C. F., Mizuno D., Non-Gaussian athermal fluctuations in active gels, *Soft Matter*, **7**, 3234-9 (2011).

[22] Valentine M. T., Kaplan P. D., Thota D., Crocker J. C., Gisler T. and Prudhomme R. K., Beck M. and Weitz D. A., Investigating the micro environments of inhomogeneous soft materials with multiple particle tracking, *Physical Review E*, **64**, 061506 (2001).

[23] Metzler R., Jeon J. H., Cherstvy A. G., Non-Brownian diffusion in lipid membranes: Experiments and simulations, *Biochimica et Biophysica Acta (BBA)-Biomembranes* **1858**, 2451-67 (2016).

[24] Metzler R., Gaussianity fair: the riddle of anomalous yet non-Gaussian diffusion, *Biophysical journal*, **112**, 413 (2018).

[25] Cherstvy A. G., Thapa S., Wagner C. E. and Metzler R., Non-Gaussian, non-ergodic, and non-Fickian diffusion of tracers in mucin hydrogels, *Soft Matter*, **15**, 2526-51 (2019).

[26] Sadoon A. A., Wang Y. Anomalous, non-Gaussian, viscoelastic, and age-dependent dynamics of histone like nucleoid-structuring proteins in live Escherichia coli. Physical Review E. 2018 Oct 19;98(4):042411.

[27] Bhowmik B. P., Tah I., Karmakar S., Non-Gaussianity of the van Hove function and dynamic-heterogeneity length scale, *Physical Review E*, **98**, 022122 (2018).

[28] Postnikov E. B., A. Chechkin and I. M. Sokolov, Brownian yet non-Gaussian diffusion in heterogeneous media: from superstatistics to homogenization, *New Journal of Physics*, (2020).

[29] Hidalgo-Soria M. and Barkai E., Hitchhiker model for Laplace diffusion processes, *Physical Review E*, **102**, 012109 (2020).

[30] Jain R. and Sebastian K. L., Diffusing diffusivity: a new derivation and comparison with simulations, *Journal of Chemical Sciences*, **129**, 929-937 (2017).

[31] Prabhakar T. R., A singular integral equation with a generalized Mittag Leffler function in the kernel, *Yokohama Math. J.*, **19**, 7-15 (1971)

[32] Podlubny I., *Fractional Differential Equations*, Academic Press, New York, 1999.

[33] Hristov J., Transient Heat Diffusion with a Non-Singular Fading Memory: From the Cattaneo Constitutive Equation with Jeffreys kernel to the Caputo-Fabrizio time-fractional derivative, *Thermal Science*, **20**, 765–770 (2016).

[34] Hristov J., *Fractional derivative with non-singular kernels: From the Caputo-Fabrizio definition and beyond: Appraising analysis with emphasis on diffusion models, In: S. Bhalekar (Ed). Frontiers in Fractional Calculus, 2017* . Bentham Science Publishers, Sharja, OAE, 2017, pp. 269-342.

[35] Hristov J., *Integral-balance solution to nonlinear subdiffusion equation*, In: S. Bhalekar (Ed). Frontiers in Fractional Calculus, 2017 . Bentham Science Publishers, Sharja, OAE, 2017, pp. 71-106.

[36] Hristov J., *Approximate solutions to time-fractional models by integral balance approach*, In: C. Cattani, H.M. Srivastava, Xia-Jun Yang, (Eds). Fractional Dynamics. De Gruyter Open, Warsaw/Berlin, 2015, pp. 78-109.

[37] dos Santos M. A. F., Fractional prabhakar derivative in diffusion equation with non-static stochastic resetting, *Physics*, **1**, 40-58 (2019).

[38] dos Santos M. A. F., Mittagleffler memory kernel in Lévy flights, *Mathematics*, **7**, 766 (2019).

[39] dos Santos M. A. F., A fractional diffusion equation with sink term, *Indian Journal of Physics*, 1-11 (2019).

[40] dos Santos M. A. F. and Gomez I. S., A fractional FokkerPlanck equation for non-singular kernel operators, *Journal of Statistical Mechanics: Theory and Experiment*, **2018**, 123205 (2018).

[41] dos Santos M. A. F., Non-Gaussian distributions to random walk in the context of memory kernels, *Fractal and Fractional*, **2**, 20 (2018).

[42] Agahi H., Khalili M., Truncated Mittag-Leffler distribution and superstatistics, *Physica A*, 124620 (2020).

[43] dos Santos M. A. F., Mittag-Leffler functions in superstatistics. *Chaos, Solitons & Fractals*, **131**, 109484 (2020).

[44] Rebenshtok A., Denisov S., Hänggi P. and Barkai E., Infinite densities for Lvy walks, *Physical Review E*, **90**, 062135 (2014).

[45] Kessler D. A. and Barkai E., Infinite covariant density for diffusion in logarithmic potentials and optical lattices, *Physical review letters*, **105**, 120602 (2010).

[46] Sposini V., Chechkin A. V., Seno F., Pagnini G. and Metzler R., Random diffusivity from stochastic equations: comparison of two models for Brownian yet non-Gaussian diffusion, *New Journal of Physics*, **20**, 043044 (2018).

[47] Sliusarenko O. Y., Vitali S., Sposini V., Paradisi P., Chechkin A., Castellani G. and Pagnini G., Finite-energy Lvy-type motion through heterogeneous ensemble of Brownian particles, *Journal of Physics A: Mathematical and Theoretical*, **52**, 095601 (2019).

[48] Hapca S., Crawford J. W. and Young I. M., Anomalous diffusion of heterogeneous populations characterized by normal diffusion at the individual level, *Journal of the Royal Society Interface*, **6**, 111-122 (2009).

[49] Budini A. A., Cáceres M O. First-passage time for superstatistical Fokker-Planck models, *Physical Review E*, **97**, 012137 (2018).

[50] Barkai E., Jung Y. and Silbey R., Theory of single-molecule spectroscopy: beyond the ensemble average, *Annu. Rev. Phys. Chem.*, **55**, 457-507 (2004).

[51] Van Der Straeten E. and Beck C., *Chinese science bulletin*, 56(34), 3633-3638 (2011).

[52] Chubynsky M. V. and Slater G. W., Diffusing diffusivity: a model for anomalous, yet Brownian, diffusion, *Physical review letters*, **113**, 098302 (2014).

In: A Closer Look at the Diffusion Equation
Editor: Jordan Hristov

ISBN: 978-1-53618-330-6
© 2020 Nova Science Publishers, Inc.

Chapter 7

MEAN SQUARE DISPLACEMENT OF THE FRACTIONAL DIFFUSION EQUATION DESCRIBED BY CAPUTO GENERALIZED FRACTIONAL DERIVATIVE

Ndolane Sene[*]
Laboratoire Lmdan, Département de Mathématiques de la Décision,
Université Cheikh Anta Diop de Dakar,
Faculté des Sciences Economiques et Gestion,
BP, Dakar Fann, Senegal

Abstract

In this chapter, we discuss the Mean Square Displacement of the fractional diffusion equation in the context of fractional calculus. The Caputo generalized fractional derivative is the main tool which we use in our study. Note that, the Mean Square Displacement is used to determine the type of diffusion processes and the analytical solution of the fractional diffusion equation. We also propose a method for getting the approximate value of the Mean Square Displacement. We illustrated our main results by depicting the graphics of the solution and the Mean Square Displacement of the fractional diffusion equation. A practical example is given to illustrate the method for getting the MSD.

---

[*]Corresponding Author's Email: ndolanesene@yahoo.fr.

Keywords: diffusion equation, mean square displacement, Caputo generalized fractional derivative

1. Introduction

Fractional calculus is a new attraction in mathematics. It is now proved fractional calculus plays an essential role in modeling the world phenomena, see in[11, 27, 34]. The fractional derivatives play an important role in this new field. But as we remark, in fractional calculus, there exist many types of fractional derivatives. Derivatives with singular kernels as the Caputo-Liouville fractional derivative[18, 29], Riemann-Liousville fractional derivative[18, 29], conformable derivative[23], Hilfer derivative[18] and many others. Derivatives with non-singular kernels as the Caputo-Fabrizio derivative [7] and the Atangana-Baleanu derivative [5, 6] are introduced in literature to avoid the singularity. In our chapter, we will use the generalized form of the Caputo-Liouville fractional derivative introduced in [19].

The applications of the fractional calculus in physics received by investigations. Fractional diffusion equation is a physical model and is the subject of intense researches in this chapter. Many types of fractional diffusion equations were under investigation in these last decade. In [34], the author tries to propose the analytical solution of the fractional diffusion equation by combining the Fourier transformation and the Laplace transform, see also in [28]. In [37], the author analyzes the impact of the Mean Square Displacement on the fractional diffusion equation described by the Mittag-Leffler fractional derivative. In [10], the author investigates on the Mean Square Displacement and proposes the solution of the fractional diffusion equation described by Prabhakar derivative. In [13], the same author gives a review of the statistical and probability properties of the fractional diffusion equations. He characterizes at the same time the sub-diffusion process, the super diffuse process, the hyper-diffusion process, the ballistic diffusion process generated by the fractional derivatives in the diffusion equations. Many other investigations related to the numerical solutions of the fractional diffusion equations exist, too, see in [8, 16, 17, 22, 20, 27, 26]. See others applications in physics in [12, 14, 25, 29, 39].

In our chapter, we investigate the Mean square Displacement of the fractional diffusion equation described by the Caputo-Liouville generalized fractional derivative. The main novelty of this work is the connection between the fractional calculus and the probability. We propose investigations on the fractional diffusion equations under Direac boundary conditions. The obtained solutions correspond the probability densities. Here, we express the density of probability using the Mean Square Displacement, and we consider a series of Random variables in which the densities are the solutions of the proposed fractional diffusion equations. We try using the data to determine the Mean Square Displacement. Also, we try to discuss the sub-diffusion, the super-diffusion, the ballistic diffusion, the hyper-diffusion, and the Richardson diffusion generated by the fractional diffusion equation with Caputo-Liouville generalized fractional derivative. The analytical solutions obtained from the fractional differential equations were graphically represented.

The chapter is structured as follows: In Section 2, we recall the fractional derivative operators. In Section 3, we discuss the qualitative properties of the head equation considered in our study. In Section 4, we propose the analytical solution of the fractional diffusion equation in terms of Fox H function. In Section 5, we propose the Mean Square Displacement of the fractional diffusion equation described by the Caputo-Liouville generalized fractional derivative. In Section 6, a novel method for getting the Mean Square Displacement has been proposed. In Section 7, we propose the analysis and interpretations of our main results. In Section 8, we finish by giving the concluding remarks and perspectives for future investigations.

2. FRACTIONAL DERIVATIVE OPERATORS AND THEIR PROPERTIES

This section addresses the definitions and properties of the fractional derivatives used in fractional calculus. We particularly recall the definitions of the generalized form of the Caputo-Liouville [18] and the Riemann-Liouville fractional derivatives [18]. Let's the function represented by $f : [b, +\infty[\longrightarrow \mathbb{R}$, we define the fractional integral of order α, for the function f starting at a as the following form [1]

$$(I^\alpha f)(t) = (I^{\alpha,1} f)(t) = \frac{1}{\Gamma(\alpha)} \int_b^t (t-s)^{\alpha-1} f(s) ds, \qquad (1)$$

where $\Gamma(...)$ is the Gamma function and the time $t > b$, with the order $0 < \alpha < 1$.

The possible generalization of the previous integral is introduced in [18, 19] and can be represented as the following form. Let's the function represented by

$f : [b, +\infty[\longrightarrow \mathbb{R}$, we generalize the integral of order $\alpha, \rho > 0$ of the function f starting at b as follows

$$I^{\alpha,\rho} f(t) = \frac{1}{\Gamma(\alpha)} \int_b^t \left(\frac{t^\rho - s^\rho}{\rho} \right)^{\alpha-1} f(s) \frac{ds}{s^{1-\rho}}, \qquad (2)$$

where $\Gamma(...)$ is the Gamma function and the time $t > 0$, with the order $0 < \alpha < 1$. We defined the fractional derivative with singular kernels. The first is the Riemann-Liouville fractional derivative [24]. Let's the function defined by $f : [b, +\infty[\longrightarrow \mathbb{R}$, we define the Riemann-Liouville fractional derivative of order α, for the function f starting at b as the following form

$$D^\alpha f(t) = \frac{1}{\Gamma(1-\alpha)} \left(\frac{d}{dt} \right) \int_b^t (t-s)^{-\alpha} f(s) ds, \qquad (3)$$

where $\Gamma(...)$ denotes the Gamma function and the time $t > b$, with the order satisfying $0 < \alpha < 1$.

Its possible generalization is introduced in [18, 19] and is defined as the following definition. Let's the function defined by $f : [b, +\infty[\longrightarrow \mathbb{R}$, we define the generalized Riemann-Liouville fractional derivative of order $\alpha, \rho > 0$ for the function f starting at b as follows

$$D^{\alpha,\rho} f(t) = \frac{1}{\Gamma(1-\alpha)} \left(t^{1-\rho} \frac{d}{dt} \right) \int_b^t \left(\frac{t^\rho - s^\rho}{\rho} \right)^{-\alpha} f(s) \frac{ds}{s^{1-\rho}}, \qquad (4)$$

where $\Gamma(...)$ is the Gamma function and time $t > 0$, with the order satisfying the condition $0 < \alpha < 1$.

Let's define the Caputo-Liouville fractional derivative [24]. We suppose the function defined by $f : [b, +\infty[\longrightarrow \mathbb{R}$, the Caputo-Liouville fractional derivative of order α, for the function f starting at the point b as the following form

$$D_c^\alpha f(t) = \frac{1}{\Gamma(1-\alpha)} \int_b^t (t-s)^{-\alpha} f'(s) ds, \qquad (5)$$

where $\Gamma(...)$ represents the Gamma function and the time $t > b$, with the order satisfying the condition $0 < \alpha < 1$.

A possible generalization were introduced in [18, 19]. Let's the function represented by $f : [b, +\infty[\longrightarrow \mathbb{R}$, the generalized Caputo-Liouville fractional derivative of order α, $\rho > 0$ for the function f starting at b can be generalized in the following form

$$D_c^{\alpha,\rho} f(t) = \frac{1}{\Gamma(1-\alpha)} \int_b^t \left(\frac{t^\rho - s^\rho}{\rho}\right)^{-\alpha} f'(s) ds, \qquad (6)$$

where $\Gamma(...)$ is the Gamma function and the time $t > b$, with the order satisfying the condition $0 < \alpha < 1$.

We define the Laplace transform related to the Caputo-Liouville generalized fractional derivative which we will use latter. The ρ-Laplace transform of the Caputo-Liouville fractional derivative can be represented for the order $0 < \alpha < 1$ by the following relationship [18, 19, 33]

$$\mathcal{L}_\rho\{D_c^{\alpha,\rho} f(t)\} = s^\alpha \mathcal{L}_\rho\{f(t)\} - s^{\alpha-1} f(b), \qquad (7)$$

where we defined the ρ-Laplace transform of the function f as the follows

$$\mathcal{L}_\rho\{f(t)\}(s) = \int_b^\infty e^{-s\frac{t^\rho}{\rho}} f(t) \frac{dt}{t^{1-\rho}}. \qquad (8)$$

The Mittag-Leffler function plays an essential role in fractional calculus, notably in the representation of the solutions of the fractional differential equations. We define the MittagLeffler function [3, 31, 38, 39] with the parameters α and β as the following series

$$E_{\alpha,\beta}(z) = \sum_{k=0}^\infty \frac{z^k}{\Gamma(\alpha k + \beta)}, \qquad (9)$$

where $\alpha > 0$, $\beta \in \mathbb{R}$ and $z \in \mathbb{C}$.

We finish by recalling the definition of the Fox H function [13, 15]. The Fox H function used in statistic can be defined as the following form

$$H_{p,q}^{m,n}\left[x\Big|_{b_q,B_q}^{a_p,A_p}\right] = \frac{1}{2\pi i} \int_L \chi(\xi) x^{-\xi} d\xi, \qquad (10)$$

where

$$\chi(\xi) = \frac{\prod_{j=1}^m \Gamma(b_j - B_j \xi) \prod_{j=1}^n \Gamma(1 - a_j + A_j \xi)}{\prod_{j=m+1}^q \Gamma(1 - b_j + B_j \xi) \prod_{j=n+1}^p \Gamma(a_j - A_j \xi)}, \qquad (11)$$

m, n, p, and q are integers and satisfy the condition $0 \leq n \leq p$ and $0 \leq m \leq q$. See more details related to this function in [13, 15].

3. QUALITATIVE PROPERTIES FOR THE FRACTIONAL DIFFUSION EQUATION

The qualitative properties for the fractional diffusion equation described by the Caputo-Liouville generalized fractional derivative is addressed in this section. This step is essential for each models introduced in fractional calculus. When we are ensured the solution of a given fractional model exists, we will try to adopt a specific method to get its solution. The fractional equation under consideration is given by the following equation

$$D_t^{\alpha,\rho} u = \nu \frac{\partial^2 u}{\partial x^2}. \tag{12}$$

In our problem, we make the following assumption, let's that

$$\psi(x,u) = \nu \frac{\partial^2 u}{\partial x^2}. \tag{13}$$

The procedure adopted in this section is called Picard's method. It consists of using the Banach fixed point Theorem. We describe the technique in the following paragraphs. Note that the method used in this section is not new and exists for the other types of fractional derivatives.

The first step consists of proving the Lipschitz continuous and determine at the same time the Lipschitz constant of the function ψ. Note that the functions u, and v are bounded, from which we can find k such that, we have the following relationship

$$\|\psi(x,u) - \psi(x,v)\| \leq k\|u - v\|. \tag{14}$$

Thus, the function ψ is Lipschitz continuous with a Lipschitz constant k. The next step of proof consists of applying the generalized fractional integral to both sides of the fractional diffusion equation Eq. (12). Its satisfies the following relationship

$$u(x,t) - u(x,0) = I^{\alpha,\rho}\psi(x,u). \tag{15}$$

Let's the Picard's operator $Ru : H \to H$ defined by the following form, that is

$$Ru(x,t) = I^{\alpha,\rho}\psi(x,u). \tag{16}$$

We will prove the function Γ is well defined. Using the Euclidean norm, we have the following relationships

$$\begin{aligned}
\|Ru(x,t) - u(x,0)\| &= \|I^{\alpha,\rho}\psi(x,u)\|, \\
&\leq I^{\alpha,\rho}\|\psi(x,u)\|, \\
&\leq \frac{\rho^{1-\alpha}}{\Gamma(\alpha)}\|\psi(x,u)\|\int_0^t \left(\frac{t^\rho - s^\rho}{\rho}\right)^{\alpha-1}\frac{ds}{s^{1-\rho}}. \quad (17)
\end{aligned}$$

From the Lipschitz continuous (Eq. (14)), we known the following fact $\|\psi(x,u)\| \leq M$ is held, and furthermore we use $t \leq T$, Eq. (17) can be represented in the following form

$$\|Ru(x,t) - u(x,0)\| \leq \frac{\rho^{1-\alpha}}{\Gamma(\alpha)}\left(\frac{T^\rho}{\rho}\right)^\alpha M. \quad (18)$$

From which we conclude, the function R is well defined. The next step consists of providing a condition under which the operator R defines a contraction. We evaluate the relationship

$$\begin{aligned}
\|Ru(x,t) - Rv(x,t)\| &= \|I^{\alpha,\rho}(\psi(x,u) - \psi(x,v))\|, \\
&\leq I^{\alpha,\rho}\|(\psi(x,u) - \psi(x,v))\|, \\
&\leq \frac{\rho^{1-\alpha}}{\Gamma(\alpha)}\|\psi(x,u) - \psi(x,v)\|\int_0^t \left(\frac{t^\rho - s^\rho}{\rho}\right)^{\alpha-1}\frac{ds}{s^{1-\rho}}.
\end{aligned}$$

Using the fact proved in Eq (14), that is the fact the function ψ is Lipschitz continuous with a Lipschitz constant k, we deduce the relationship

$$\|Ru(x,t) - Rv(x,t)\| \leq \frac{\rho^{1-\alpha}}{\Gamma(\alpha+1)}\left(\frac{T^\rho}{\rho}\right)^\alpha k\|u-v\|. \quad (19)$$

From which, the operator R defines a contraction when the following condition is held

$$\frac{\rho^{1-\alpha}}{\Gamma(\alpha+1)}\left(\frac{T^\rho}{\rho}\right)^\alpha k < 1. \quad (20)$$

Recalling the Banach fixed Theorem, the solution of the fractional diffusion equation described by the Caputo-Liouville generalized fractional derivative (12) exists. The problem of investigating the analytical solution or providing the approximate solution is well defined.

4. ANALYTICAL SOLUTION OF THE FRACTIONAL DIFFUSION EQUATION

In this section, we discuss the analytical solution of the generalized fractional diffusion equation described by the generalized fractional derivative using the Fourier transformation. We will use the analytical solution to determine the mean square displacement (MSD) associated to the generalized fractional diffusion equation. We give some consequences generated by the MSD in physics. The following equation provides the equation under consideration

$$D_t^{\alpha,\rho} u = \frac{\partial^2 u}{\partial x^2}. \tag{21}$$

We add the statistical boundary condition given by

$$u(x,0) = \delta(x). \tag{22}$$

where δ denotes the Dirac function. The used boundary condition is the new attraction for the generalized fractional diffusion equations. Our investigations are to discuss the meaning of the proposed model physically in detail in the context of fractional calculus. It follows from Fourier transformation and Laplace transform the following identity

$$\begin{aligned} s^\alpha \tilde{u}(q,s) - s^{\alpha-1} &= -q^2 \tilde{u}(q,s), \\ \tilde{u}(q,s) &= \frac{s^{\alpha-1}}{s^\alpha + q^2}. \end{aligned} \tag{23}$$

Applying the inverse of the Fourier transform and at the same time the inverse of the Laplace transform, we obtain the following analytical solution

$$u(x,t) = \frac{1}{\sqrt{4\left(\frac{t^\rho}{\rho}\right)^\alpha}} H_{1,1}^{1,0}\left[\frac{|x|}{\sqrt{\left(\frac{t^\rho}{\rho}\right)^\alpha}} \bigg|_{0,1}^{1-\frac{\alpha}{2},\alpha}\right]. \tag{24}$$

In the next section, we provide the Mean Square Displacement of the generalized fractional diffusion equation by using the Fourier Laplace transform established in Eq. (23). The Fourier and Laplace transformation plays a fundamental role in the problem consisting of getting the Mean Square Displacement, as we will notice in the next section.

5. MEAN SQUARE DISPLACEMENT FOR FRACTIONAL DIFFUSION EQUATION

In this section, we address the Mean Square Displacement of the generalized fractional diffusion equation (21) described by the Caputo-Liouville generalized fractional derivative. It is known that there exists a linear relationship between the Mean Square Displacement and the time for the classical diffusion equations obtained when $\alpha = \rho = 1$. What about in fractional calculus, in the case of the Caputo–Liouville generalized fractional derivative. One of the objective in this section consists of getting the relationship between the Mean Square Displacement and the diffusion coefficient ν. The MSD is essential in physics because it gives the nature of diffusion processes: normal diffusion, sub-diffusion, ballistic diffusion, and others. Note that in the physics, there exist a technique of determination of the solution using the form of the similarity variable. This chapter contributes too to give the structure of the similarity variable for the generalized fractional diffusion equation and give its relation with the diffusion coefficients. The following relationships provide the value of the Mean Square Displacement in the context of Fourier transformation

$$MSD = \lim_{q \to 0} -\frac{d^2 \tilde{u}}{dq^2}, \qquad (25)$$

where the function $\tilde{u}$ is the Fourier and the Laplace transformations obtained in Eq. (23). In the way to get the Mean Square Displacement, we first determine the function $\frac{d\tilde{u}}{dq}$, we have the following calculation

$$\frac{d\tilde{u}}{dq} = -\frac{2q\nu s^{\alpha-1}}{(s^\alpha + q^2 \nu)^2}. \qquad (26)$$

The second derivative is given by the following equation

$$\frac{d^2 \tilde{u}}{dq^2} = -\frac{2\nu}{s^{\alpha+1}}. \qquad (27)$$

Applying the inverse of the Laplace transform to both sides of Eq. (27), we get the same Mean Square Displacement as in Stokes equation, we have

$$MSD = \frac{2\nu}{\Gamma(1+\alpha)} \left(\frac{t^\rho}{\rho}\right)^\alpha. \qquad (28)$$

A good result can be deduced related to the analytical solution. Our novelty in this chapter is the method to get an analytical solution. The procedure is described as follows: in the first step, we get the Mean Square Displacement. In this second step, we propose the solution of the fractional diffusion equation by using the similarity variable. We have the following solution form

$$u(x,t) = \frac{1}{\sqrt{2MSD}} H_{1,1}^{1,0}\left[\frac{|x|}{\sqrt{MSD/2}}\Big|_{0,1}^{1-\frac{\alpha}{2},\alpha}\right], \qquad (29)$$

where the similarity variable is denoted by $x/\sqrt{MSD}$. The similarity surface will be represented in Section 7. When $\alpha = 1$ and ρ is arbitrary, the solution of the fractional diffusion equation can be represented with Gaussian function. After translating the Fox H function, the solution of the generalized diffusion equation can be represented as follow

$$u(x,t) = 1 - erf\left(\frac{x}{\sqrt{MSD/2}}\right) = 1 - erf\left(\frac{x}{\left(\frac{t^\rho}{\rho}\right)}\right), \qquad (30)$$

where $erf(...)$ represents the Gaussian error function. At last, we can deduce the diffusion coefficient ν using the deviation of the particle expressed in the following form

$$\nu = \frac{\Gamma(1+\alpha)MSD}{\left(\frac{t^\rho}{\rho}\right)^\alpha}. \qquad (31)$$

In Section 7, we depict the behavior of the solutions represented in Eq. (30), the Mean Square Displacement, and discuss the natures of the diffusion processes.

6. Method for Getting the Mean Square Displacement

In this section, we propose a novel method for getting the Mean Square Displacement. Let's a series of random variables X_i for $i = 1, 2, ..., n$ which have the same density functions defined by the solution of the fractional diffusion equation (21) under Eq. (22). Let's $\alpha = 1$, we define the density as the form

$$u(x_i, t) = \frac{1}{\sqrt{2\pi MSD}} \exp\left(-\frac{x_i^2}{2MSD}\right). \qquad (32)$$

Note the solution represented in Eq. (32) is obtained after transforming the solution represented by the Gaussian error function in Eq. (30). In Probability, there exists a method (maximum likelihood) for estimating an unknown parameter when we consider a series of random variables. This method is applied to our problem. The first step consists of determining the function L defined as follows

$$L(MSD) = \prod_{i=1}^{n} u(x_i, t) = \prod_{i=1}^{n} \frac{1}{\sqrt{2\pi MSD}} \exp\left(-\frac{x_i^2}{2MSD}\right). \qquad (33)$$

The second step consists of getting the function $\ln(L)$ defined as the form

$$\ln(L(MSD)) = \sum_{i=1}^{n} u(x_i, t) = \sum_{i=1}^{n} \ln\left[\frac{1}{\sqrt{2\pi MSD}} \exp\left(-\frac{x_i^2}{2MSD}\right)\right]. \qquad (34)$$

After calculation omitted due to space limitation, the value of the function $\ln(L)$ is given by the relationship

$$\ln L(MSD) = -\frac{n}{2}\ln(2\pi) - \frac{n}{2}\ln(MSD) - \frac{\sum_{i=1}^{n} x_i^2}{2MSD}. \qquad (35)$$

Calculating the derivative of the function $\ln(L)$ regarding the unknown parameter MSD, we get

$$\frac{\partial \ln L}{\partial MSD} = 0 \implies \frac{-n}{2MSD} + \frac{\sum_{i=1}^{n} x_i^2}{2MSD^2}. \qquad (36)$$

Finally, the approximate value of the MSD for the fractional diffusion equation described by the Caputo-Liouville generalized fractional derivative is given by the relationship

$$MSD = \frac{\sum_{i=1}^{n} x_i^2}{n}. \qquad (37)$$

One can verify this value is the maximum value for the MSD because

$$\frac{\partial^2 \ln L}{\partial MSD^2} = -\frac{n}{2MSD^2} < 0. \qquad (38)$$

Our problem constitutes a new advance in fractional calculus because when the generalized fractional derivative is used, and $\alpha = 1$, the fractional-order ρ can be determined by using the data. The formula given the order using data is provided by

$$\frac{t^\rho}{\rho} = \frac{\sum_{i=1}^{n} x_i^2}{n}. \qquad (39)$$

Let's give a simple example of the determination of the Mean Square Displacement MSD for the fractional diffusion equation (12). Our paper is a direct application of fractional calculus in real-life problems. We let a series defined by

i	1	2	3	4	5	6	7	8
x_i	0.54	0.53	0.59	0.66	0.63	0.62	0.65	0.60
x_i^2	0.2916	0.2809	0.3481	0.4356	0.3969	0.3844	0.4225	0.3600

Using Eq. (37), the value of the Mean Square Displacement is given by

$$MSD = 0.365. \qquad (40)$$

7. ANALYSIS AND INTERPRETATIONS OF THE MAIN RESULTS

In this section, we did some graphical representations and interpretations. At first, let's analyze the nature of the diffusion process generated by the fractional equation (12). The types of diffusion processes, we can cite normal diffusion, super diffusion, ballistic diffusion, Richardson diffusive process, and others. All these diffusion processes correspond to a specific value of fractional order, which we will try to determine in this section using the formula of the MSD. From the formulation of the Mean Square Displacement in Eq. (28); the normal diffusion is obtained when

$$\alpha\rho = 1 \implies \rho = \frac{1}{\alpha} \qquad (41)$$

In Figure 1a, we depict some normal diffusion processes for different values of the orders $\alpha \in (0, 1)$. We remark when the order $\rho = 1/\alpha$ the MSD in Eq. (28) is linear in time, we recover the classical behavior of the particle deviation. Thus the order ρ plays a fundamental role in the diffusion processes. It can be considered as a regulator. We note for increasing order of $\alpha \in (0, 1)$, we note the MSD increase and converge to the MSD obtained in the classical case with the values ($\rho = \alpha = 1$). In conclusion, when the order ρ converges to infinity,

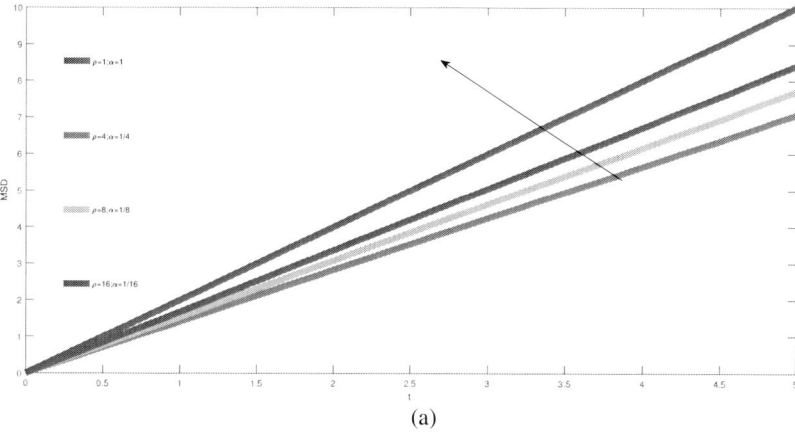

Figure 1. MSD for normal diffusion with different orders of $\rho = 1, 4, 8, 16$.

we note the diffusion process converges to the classical diffusion. But we note a retardation effect generated by the parameter ρ. The fractional equation (21) equation generate a sub-diffusion equation when

$$\alpha\rho < 1 \Longrightarrow \alpha < \frac{1}{\rho}, \qquad (42)$$

The condition (42) is possible if and only if $\rho > 1$. Note that the condition $\rho < 1/\alpha$ which is possible can not be considered in our study due to the fact $\alpha \in (0, 1)$. In Figure 2a, we depict the MSD for the sub-diffusion process for different values of α and ρ respecting the Eq. (42). We fix $\rho = 2$, we notice when the order α increases in into $(0, 1)$, the MSD in sub-diffusive process increases as well and converges slowly to the classical MSD ($\rho = \alpha = 1$). We also notice the MSD of the sub-diffusion process is not linear in time, but parabolic (see arrow). In general, we note a retardation effect generated by the parameter ρ. The fractional equation (21) generates a super-diffusive process when

$$1 < \alpha\rho < 2 \Longrightarrow \frac{1}{\alpha} < \rho < \frac{2}{\alpha}. \qquad (43)$$

In Figure 3a, we depict the MSD when the fractional diffusion equation (21) generates a super-diffusive process for different orders respecting the condition established in Eq. (43) and we fix $\alpha = 1$. We note for super-diffusive process generated by fractional equation (21) when the order ρ increases, the MSD is

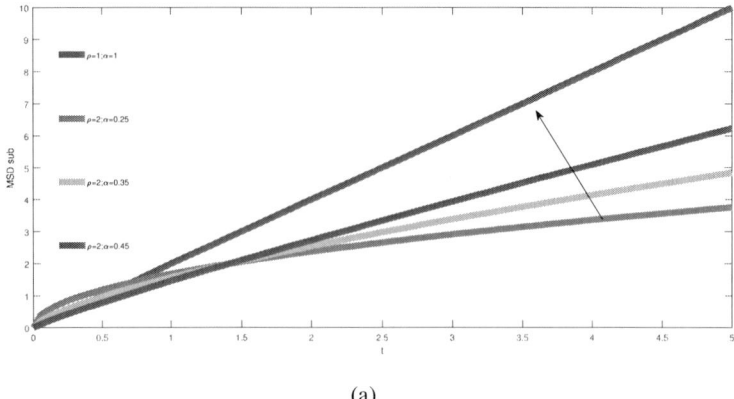

(a)

Figure 2. MSD for sub-diffusion with different orders of $\alpha = 1, 0.25, 0.35, 0.45$.

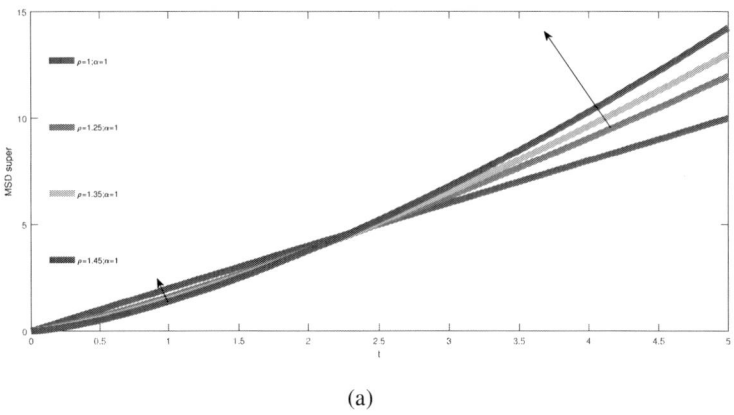

(a)

Figure 3. MSD for super-diffusion with different orders of $\rho = 1, 1.25, 1.35, 1.45$.

not linear in time and do not converge to the normal diffusion ($\rho = \alpha = 1$). The ballistic diffusion process generated by the fractional diffusion equation (21) can be recovered in our study when

$$\alpha\rho = 2 \implies \rho = \frac{2}{\alpha} \qquad (44)$$

Mean Square Displacement of the Fractional Diffusion Equation ...

The MSD of this type of diffusion can be observed in Figure 4. We observe

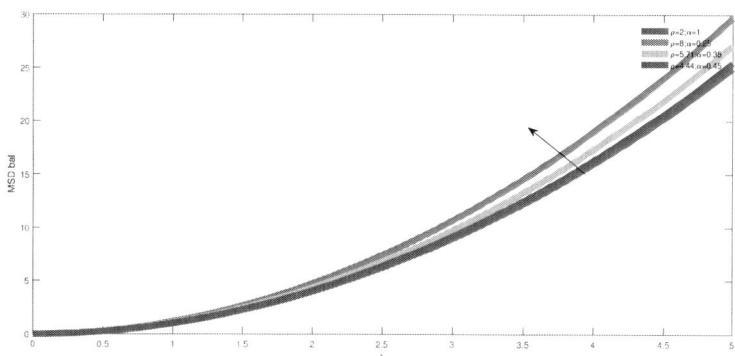

Figure 4. MSD for ballistic diffusion with different orders.

when the order α increase then the MSD increases rapidly as well but no linear in time, see arrow in Figure 4. The hyper-diffusive process is generated by the fractional diffusion equation (21) when

$$2 < \alpha\rho < 3 \Longrightarrow \frac{2}{\alpha} < \rho < \frac{3}{\alpha}. \qquad (45)$$

And we depict in Figure 5a, the MSD of the hyper diffusive process for different values of the orders α and ρ respecting the Eq. (45). We notice when the order α increases, the deviation of the particle decreases or increases. The Richardson diffusion process generated by the fractional diffusion equation can be recovered in our study when

$$\alpha\rho = 3 \Longrightarrow \rho = \frac{3}{\alpha} \qquad (46)$$

We depict in figure 6a, the Richardson diffusion process. The impact of the order ρ has an acceleration effect when its value increases. We finish by depicting in the same figure all diffusion processes generated by the fractional diffusion equation (21). We have fixed all $\alpha = 1$. We notice all deviations converge after a specific time to the normal diffusion process. The order ρ plays a fundamental role when you want to control the type of diffusion process. Note that, the order α into $(0, 1)$ habitually generate a sub-diffusion. The other types of diffusion processes are possible when the fractional derivative is bi-orders, as in our

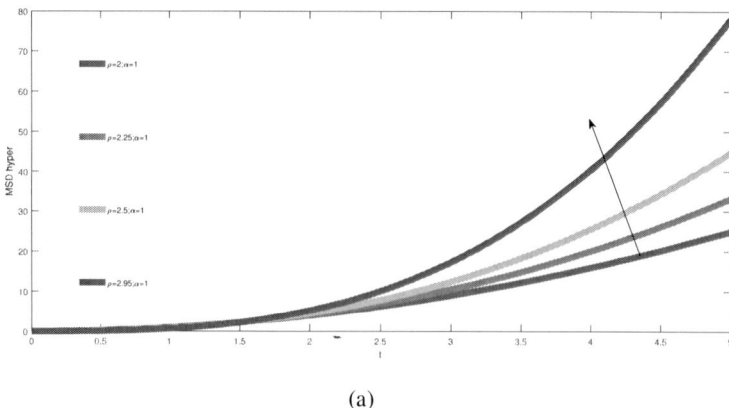

Figure 5. MSD for Hyper-diffusion with different order of $\rho = 2, 2.25, 2.5, 2.95$.

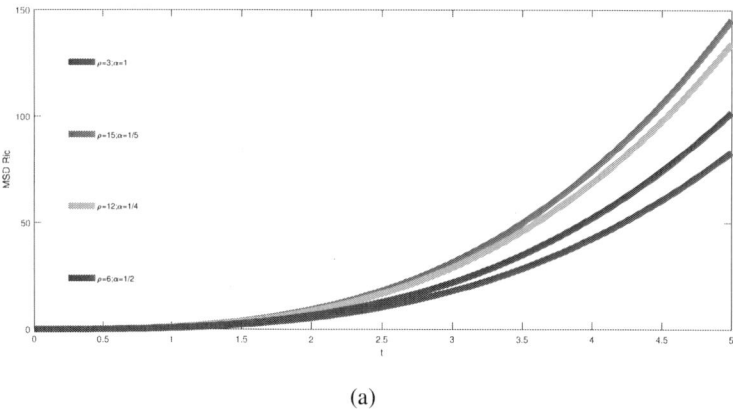

Figure 6. MSD for Richardson diffusion with different order of $\rho = 3, 6, 8, 15$.

investigation. In conclusion, with the Caputo-Liouville generalized fractional derivative, all types of diffusion processes are possible. It depends on the values of the order ρ. And the order ρ has acceleration or a retardation effect in the diffusion process.

Let's now analyze and depict graphically the probability density generated by the solution of the fractional diffusion equation (21). In this section, we

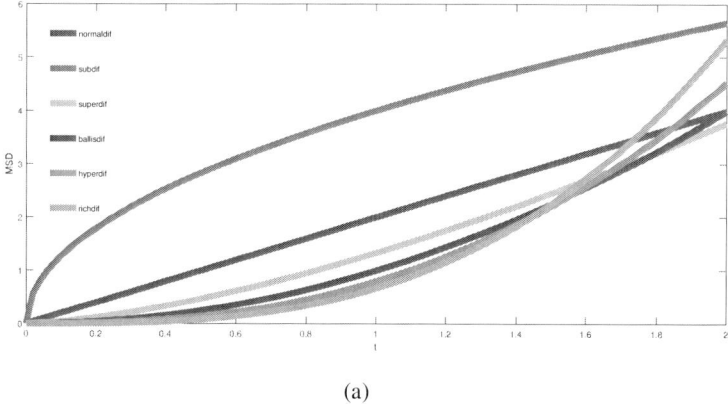

Figure 7. MSD for different types of diffusion processes.

suppose $\alpha = 1$ and done analysis regarding the values of the order ρ. Note that in Eq. (29), when $\alpha = 1$, the solution can be represented as the following form

$$u(x,t) = \frac{1}{\sqrt{4\pi \left(\frac{t^\rho}{\rho}\right)}} \exp\left(-\frac{x^2}{4\left(\frac{t^\rho}{\rho}\right)}\right) \tag{47}$$

In figure 8a, we depict for increasing MSD, the Gaussian profiles obtained with the fractional diffusion equation described by the Caputo-Liouville generalized fractional derivative. We note when the MSD increases, the probability density decreases as well, and the Pearson number does not exceed the characteristic value 3. Note that the Pearson number measures the applatisment of the Gaussian profiles.

As asked in the previous section, we depict the similarity surface in Figure 9a when the order $\alpha = \rho = 0.5$. To highlight the behavior of the similarities variables, we fix $\alpha = 1$, $x = 1$ and we depict the similarities variables for the sub-diffusive process for an increasing order ρ, see in Figure 10a. We note in Figure 10a when the order ρ increases, the similarity variable decreases rapidly and converge to zero.

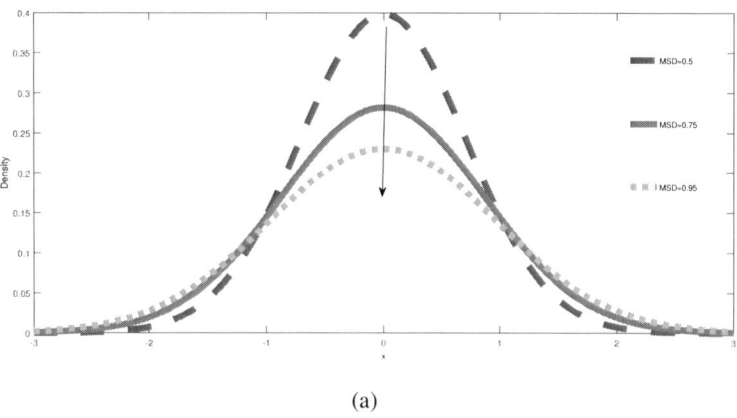

(a)

Figure 8. Probability density for $\alpha = 1$.

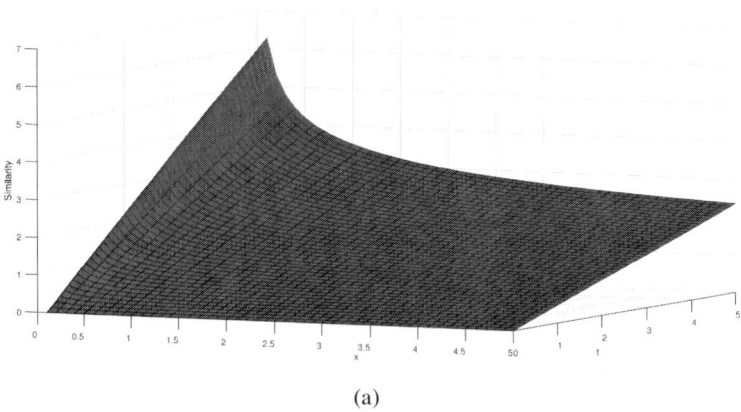

(a)

Figure 9. Similarity surfaces $\alpha = \rho = 0.5$.

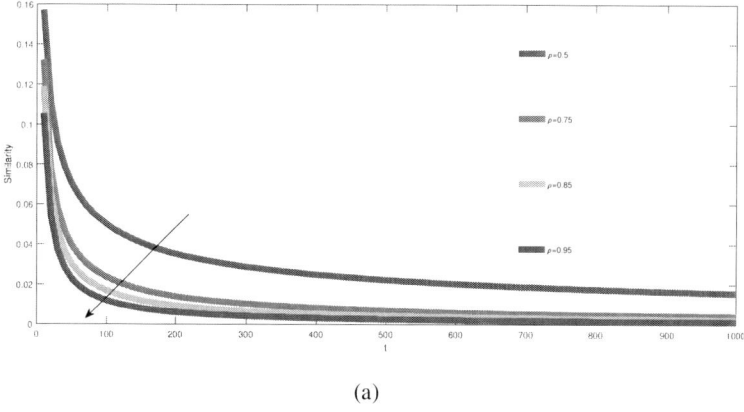

(a)

Figure 10. Similarities lines $\alpha = 1$ for increasing ρ.

Conclusion

In this chapter, we have discussed the type of diffusion processes generated by the fractional diffusion equation with the Caputo-Liouville generalized fractional derivative. We have proposed a novel method for getting the Mean Square Displacement when the solution of the proposed fractional diffusion equation is a density of probability for a random variable. This investigation is a direct application of fractional calculus in a real word problem and opens a new door for the use of data in fractional calculus investigations. The Mean Square Displacement for normal diffusion, sub-diffusion, super-diffusion, hyper-diffusion, ballistic diffusion, and Richardson diffusion are represented graphically to observe their behaviors. Investigating the diffusion process when the boundary condition is the Dirac function is an open problem, and the form of the solution can be developed and generalized in future works.

References

[1] Alkahtani B. S. T., Atangana-Batogna numerical scheme applied on a linear and non-linear fractional differential equation, *Euro. Phys. J. Plus*, **133(3)**, 111, (2018), doi: 10.1140/epjp/i2018-11961-8.

[2] Abdeljawad T., Al-Mdallal Q. M., Discrete Mittag-Leffler kernel type fractional difference initial value problems and Gronwall's inequality, *J. Comput. Appl. Math.*, **339**, 218-230 (2018).

[3] Abdeljawad T., Baleanu D., On fractional derivatives with generalized Mittag-Leffler kernels, *Adv. Diff. Equat.*, **468**, (2018), doi: 10.1186/s13662-018-1914-2.

[4] Atangana A. and Baleanu D., New fractional derivatives with nonlocal and non-singular kernel: theory and application to heat transfer model, *Thermal Sci.*, **20(2)**, 763-769 (2016).

[5] Atangana A., and Koca I., Chaos in a simple nonlinear system with AtanganaBaleanu derivatives with fractional order, *Chaos, Soli. and Fract.*, **89**, 447-454 (2016).

[6] Atangana A., Owolabi K. M., New numerical approach for fractional differential equations, *Math. Mod. Nat. Phen.*, **13(1)**, 3, (2018), doi: 10.1051/mmnp/2018010.

[7] Caputo M., and Fabrizio M., A new definition of fractional derivative without singular kernel, *Progr. Fract. Differ. Appl.*, **1(2)**, 1-15 (2015).

[8] Damor R. S., Kumar S., and Shukla A. K., Numerical Solution of Fractional Diffusion Equation Model for reezing in Finite Media, *Int. J. Engin. Math.*, 8, 2013, doi: 10.1155/2013/785609.

[9] Santos M. D., Gomez I. S., A fractional Fokker-Planck equation for non-singular kernel operators, *J. Stat. Mech. Theory Exp.*, 123205, (2018), doi: 10.1088/1742-5468/aae5a2.

[10] Santos M. D., Fractional Prabhakar Derivative in Diffusion Equation with Non-Static Stochastic Resetting, *Phys.*, **1**, 40-58 (2019).

[11] Santos M. D., Non-Gaussian Distributions to Random Walk in the Context of Memory Kernels, *Fractal Fract.*, **2**, 20, (2018), doi: 10.3390/fractalfract2030020.

[12] Santos M. D., From continuous-time random walks to controlled-diffusion reaction, *J. Stat. Mech.*, (2019), doi: 10.1088/1742-5468/ab081b.

[13] Santos M. D., Analytic approaches of the anomalous diffusion: A review, *Chaos, Soli. and Fract.*, **124**, 86-96 (2019).

[14] Fazio R., Jannelli A., and Agreste S., A Finite Difference Method on Non-Uniform Meshes for Time-Fractional Advection-Diffusion Equations with a Source Term, *Appl. Sci.*, **8**, 960, (2018), doi: 10.3390/app8060960.

[15] Glöckle W. G., Nonnenmacher T. F, Fox function representation of non-debye relaxation processes, *J. Stat. Phys.*, **71(3-6)**, 741-757 (1993).

[16] Hashemi M. S., Baleanu D., Haghighi M. P., Solving the time fractional diffusion equation using a lie group integrator, *Therm. Scien.*, **19**, 77-83 (2015).

[17] Hanert E., Piret C., Numerical solution of the space-time fractional diffusion equation: Alternatives to finite differences, *5th IFAC Symp. Fract. Diff. Appli. FDA2012*, (2012), http://hdl.handle.net/2078.1/111144.

[18] Fahd J. and Abdeljawad T., A modified Laplace transform for certain generalized fractional operators, *Res. Nonl. Anal.*, **2**, 88-98 (2018).

[19] Fahd J., Abdeljawad T., Generalized fractional derivatives and Laplace transform, *Disc. Cont. Dyn. Sys. s*, 1775-1786 (2019).

[20] Meerschaert M. M., Tadjeran C., Finite difference approximations for fractional advection-dispersion flow equations, *J. Comput. App. Math.*, **172**, 65-67 (2004).

[21] Al-Refai M., and Abdeljawad T., Analysis of the fractional diffusion equations with fractional derivative of non-singular kernel, *Adv. Diff. Equat.*, 315, (2017), doi: 10.1186/s13662-017-1356-2.

[22] Khader M. M., On the numerical solutions for the fractional diffusion equation, *Commun Nonlinear Sci Numer Simulat*, **16**, 2535-2542 (2011).

[23] Khalil R., Horani M. Al., Sababheh Y. M., A new definition of fractional derivative, *J. Comp. appl. Math.*, **264**, 65-70 (2014).

[24] Kilbas A. A., Srivastava H. M., and Trujillo J. J., Theory and Applications of Fractional Differential Equations, North-Holland Mathematics Studies, Elsevier, Amsterdam, The Netherlands, **204** (2006).

[25] Tasbozan O., Esen A., Yagmurlu N.M., and Ucar Y., A Numerical Solution to Fractional Diffusion Equation for Force-Free Case, *Abst. Appli. Anal.*, **6**, (2013), doi: 10.1155/2013/187383.

[26] Owolabi K. M., Atangana A., On the formulation of Adams-Bashforth scheme with Atangana-Baleanu-Caputo fractional derivative to model chaotic problems, *Chaos*, **29**, 023111, (2019), doi: 10.1063/1.5085490.

[27] Pimenov V. G., Hendy A. S., A numerical solution for a class of time fractional diffusion equations with delay, *Int. J. Appl. Math. Comput. Sci.*, **27(3)**, 477-488 (2017).

[28] Sene N., Analytical solutions of Hristov diffusion equations with non-singular fractional derivatives, *Chaos*, **29**, 023112, (2019), doi: 10.1063/1.5082645.

[29] Sene N., Solutions of fractional diffusion equations and Cattaneo-Hristov diffusion models, *Int. J. Appli. Anal.*, **17(2)**, 191-207 (2019).

[30] Sene N., Lyapunov characterization of the fractional nonlinear systems with exogenous input, *Frac. Fract.*, **2(2)**, (2018), doi: 10.3390/fractalfract2020017.

[31] Sene N., Homotopy Perturbation ρ-Laplace Transform Method and Its Application to the Fractional Diffusion Equation and the Fractional Diffusion-Reaction Equation, *Frac. Fract.*, **3**,14, (2019), doi: 10.3390/fractalfract3020014.

[32] Sene, N., Stokes first problem for heated flat plate with AtanganaBaleanu fractional derivative, *Chaos, Soli. Fract.*, **117**, 68-75 (2018).

[33] Fall A. N., Ndiaye S. N., Sene N., Black-Scholes option pricing equations described by the Caputo generalized fractional derivative, *Chaos, Soli. Fract.*, **125**, 108-118 (2019).

[34] Sene N., Analytical solutions and numerical schemes of certain generalized fractional diffusion models, *Eur. Phys. J. Plus*, **134**, 199, (2019), doi: 10.1140/epjp/i2019-12531-4.

[35] Sene N., Gomez-Aguilar J.F., Analytical solutions of electrical circuits considering certain generalized fractional derivatives, *Eur. Phys. J. Plus*, **134**, 260, (2019), doi: 10.1140/epjp/i2019-12618-x.

[36] Sene N., Gomez-Aguilar J.F., Fractional Mass-Spring-Damper System Described by Generalized Fractional Order Derivatives, *Fractal Fract.*, **3**, 39, (2019), doi: 10.3390/fractalfract3030039.

[37] Sene N., Abdelmalek K., Analysis of the fractional diffusion equations described by Atangana-Baleanu-Caputo fractional derivative, *Chaos, Soli. Fract.*, **127**, 158-164 (2019).

[38] Sene N., Srivastava G., Generalized Mittag-Leffler Input Stability of the Fractional Differential Equations, *Sym.*, **11**, 608, (2019), doi: 10.3390/sym11050608.

[39] Sene N., Integral Balance Methods for Stokes First, Equation Described by the Left Generalized Fractional Derivative, *Phys.*, **1**, 154-166 (2019).

EDITOR CONTACT INFORMATION

Jordan Hristov, PhD, DSc
Professor of Chemical Engineering
Department of Chemical Engineering
University of Chemical Technology and Metallurgy, Sofia
Sofia, Bulgaria
Email: jordan.hristov@mail.bg;
jordan.hristov@mail.ru;
hristovmeister@gmail.com

INDEX

A

ABR derivative, viii, 117, 120
adsorption isotherms, 23
anomalous diffusion, viii, 94, 113, 114, 115, 140, 146, 171
Atangana-Baleanu derivative in Riemann-Liouville sense (ABR), viii, 117, 119, 120

B

Binary Collision Theory (BCT) model, 27, 35, 41, 45, 46, 47, 48, 50
boundary value problem, vii, viii, 2, 5, 6, 20, 25
Brownian motion, 114, 134
Brownian particle, 136, 140, 145, 150

C

Caputo generalized fractional derivative, 151, 172
Caputo-Fabrizio derivative (CF), 119
CFD, 28, 30, 39, 50, 51
Constant Lewis Number model (CLN), 41, 43, 44, 45, 46, 47, 48

D

differential equations, 4, 114, 117, 131, 132, 170

diffusion equation, viii, ix, 1, 3, 20, 22, 23, 24, 25, 27, 28, 30, 36, 48, 50, 55, 58, 59, 62, 75, 79, 82, 86, 89, 90, 91, 93, 94, 95, 97, 99, 100, 101, 103, 104, 105, 106, 107, 109, 110, 111, 112, 113, 115, 116, 118, 119, 130, 131, 133, 134, 149, 151, 152, 153, 155, 156, 157, 158, 159, 160, 161, 162, 163, 164, 165, 166, 167, 169, 170, 171, 172, 173
diffusion process, ix, 3, 4, 15, 55, 56, 60, 68, 95, 102, 112, 113, 118, 135, 136, 138, 151, 152, 159, 160, 162, 163, 164, 165, 166, 167, 169
diffusivities, vii, ix, 58, 59, 62, 65, 133, 136, 137, 140, 145, 147
diffusivity, ix, 2, 22, 23, 34, 35, 37, 41, 45, 46, 50, 55, 56, 57, 58, 59, 60, 63, 66, 68, 72, 75, 82, 83, 84, 86, 87, 88, 89, 90, 134, 135, 136, 137, 138, 140, 141, 142, 143, 145, 148, 149, 150
Double Integration Method (DIM), viii, 55, 65, 66, 67, 68, 69, 70, 71, 72, 73, 76, 77, 78, 79, 80, 82, 84, 85, 86, 87

E

exponential diffusivity, 55, 58, 66, 68, 86, 90

F

fading memory, viii, 117
finite difference method, 2, 6, 7, 20, 23, 171
finite element method, 3
finite speed, 60, 61, 65

Fokker-Planck equation (FP), 134, 136, 140, 141, 146, 170
Fourier method, ix, 117, 129
Fourier transform, 98, 99, 107, 108, 137, 139, 141, 142, 144, 145, 152, 158, 159
fractional derivatives, 94, 130, 131, 152, 153, 156, 170, 171, 172

G

generalised distributions, 134
Green function approach, viii, 94, 95, 96, 97, 103

H

heat conduction, 23, 24, 91, 94
heat transfer, 30, 52, 94, 170
Heat-Balance Integral (HBIM), viii, 65, 66, 67, 69, 70, 71, 73, 77, 78, 87, 91
Hristov diffusion, 117, 119, 121, 125, 131, 172
hypersonic flow field, 28

J

Jacobian Free Newton Krylov (JFNK) method, 30, 40
Jacobian matrix, 7, 30, 40

K

kinetic equations, 104
kinetics, 94, 114

L

liquids, 2, 60, 146

M

mean square displacement, ix, 94, 96, 101, 106, 108, 111, 133, 134, 151, 152, 153, 155, 157, 158, 159, 160, 161, 162, 163, 165, 167, 169, 171, 173

N

Navier-Stokes Equations, 27, 30, 39, 52
nonlinear diffusion, 2, 21, 22, 24, 57, 59, 61, 63, 65, 67, 69, 71, 73, 75, 77, 79, 81, 83, 85, 87, 89, 90, 91, 95

O

optical lattice, 149
optimization, 66, 87

P

partial differential equation(s) (PDEs), 2, 5, 21, 24, 25, 90
probability density function, viii, 94, 101
probability distribution, 134

R

radiation heat transfer, 94
random environment, 137
random walk, viii, 93, 95, 97, 101, 105, 111, 114, 116, 130, 134, 146, 149, 170
Reynolds Averaged Navier-Stokes (RANS) equations, 30
Reynolds number, 34, 41
Riemann-Liouville derivative (RL), 95, 96, 100, 118, 120, 124, 130

S

Species Continuity Equations, 32, 34
Stefan-Maxwell diffusion equation (SMDE), viii, 41, 48, 50
superstatistics, ix, 134, 135, 136, 137, 139, 140, 141, 143, 145, 146, 147, 148, 149

T

thermodynamic equilibrium, 134
thermodynamic properties, 52
thermodynamics, 28
third boundary condition, 83

transformation, 61, 62, 63, 70, 84, 152, 158, 159
transport, vii, 34, 35, 53, 87, 89, 94, 113, 114, 115, 118, 121

V

Volterra integral equation, 125, 130

W

water sorption, 88
wetting, vii, 56, 60, 62, 63, 64, 65, 88